T0336068

Achieving Full Realization and Mitigating the Challenges of the Internet of Things

Marcel Ohanga Odhiambo
Mangosuthu University of Technology, South Africa

Weston Mwashita
Vaal University of Technology, South Africa

A volume in the Advances in
Wireless Technologies and
Telecommunication (AWTT) Book
Series

Published in the United States of America by
IGI Global
Engineering Science Reference (an imprint of IGI Global)
701 E. Chocolate Avenue
Hershey PA, USA 17033
Tel: 717-533-8845
Fax: 717-533-8661
E-mail: cust@igi-global.com
Web site: http://www.igi-global.com

Library of Congress Cataloging-in-Publication Data

Names: Odhiambo, Marcel, 1955- editor. | Mwashita, Weston, editor.
Title: Achieving full realization and mitigating the challenges of the
  Internet of Things / Marcel Odhiambo, and Weston Mwashita, editors.
Description: Hershey, PA : Information Science Reference, an imprint of IGI
  Global, [2022] | Includes bibliographical references and index. |
  Summary: "IoT technologies are being used in the re-engineering of a
  variety of products resulting in better performance, reduced cost, and
  an improved customer experience, and this book offers a variety of
  chapters that specifically addresses the challenges faced in the rolling
  out of IoT technologies "-- Provided by publisher.
Identifiers: LCCN 2021051921 (print) | LCCN 2021051922 (ebook) | ISBN
  9781799893127 (h/c) | ISBN 9781668448915 (s/c) | ISBN 9781799893134
  (ebook)
Subjects: LCSH: Internet of things.
Classification: LCC TK5105.8857 .A28 2022 (print) | LCC TK5105.8857
  (ebook) | DDC 004.67/8--dc23/eng/20211220
LC record available at https://lccn.loc.gov/2021051921
LC ebook record available at https://lccn.loc.gov/2021051922

This book is published in the IGI Global book series Advances in Wireless Technologies and Telecommunication (AWTT) (ISSN: 2327-3305; eISSN: 2327-3313)

British Cataloguing in Publication Data
A Cataloguing in Publication record for this book is available from the British Library.

All work contributed to this book is new, previously-unpublished material.
The views expressed in this book are those of the authors, but not necessarily of the publisher.

For electronic access to this publication, please contact: eresources@igi-global.com.

Advances in Wireless Technologies and Telecommunication (AWTT) Book Series

ISSN:2327-3305
EISSN:2327-3313

Editor-in-Chief: Xiaoge Xu University of Nottingham Ningbo China, China

MISSION

The wireless computing industry is constantly evolving, redesigning the ways in which individuals share information. Wireless technology and telecommunication remain one of the most important technologies in business organizations. The utilization of these technologies has enhanced business efficiency by enabling dynamic resources in all aspects of society.

The **Advances in Wireless Technologies and Telecommunication Book Series** aims to provide researchers and academic communities with quality research on the concepts and developments in the wireless technology fields. Developers, engineers, students, research strategists, and IT managers will find this series useful to gain insight into next generation wireless technologies and telecommunication.

COVERAGE

- Broadcasting
- Virtual Network Operations
- Wireless Technologies
- Mobile Web Services
- Digital Communication
- Mobile Communications
- Global Telecommunications
- Telecommunications
- Radio Communication
- Grid Communications

IGI Global is currently accepting manuscripts for publication within this series. To submit a proposal for a volume in this series, please contact our Acquisition Editors at Acquisitions@igi-global.com or visit: http://www.igi-global.com/publish/.

Titles in this Series

For a list of additional titles in this series, please visit:
www.igi-global.com/book-series/advances-wireless-technologies-telecommunication/73684

Modelling and Simulation of Fast-Moving Ad-Hoc Networks (FANETs and VANETs)
T.S. Pradeep Kumar (Vellore Institute of Technology, India) and M. Alamelu (B.S. Abdur Rahman University, India)
Information Science Reference • © 2022 • 300pp • H/C (ISBN: 9781668436103) • US $215.00

Handbook of Research on Design, Deployment, Automation, and Testing Strategies for 6G Mobile Core Network
D. Satish Kumar (Nehru Institute of Engineering and Technology , India) G. Prabhakar (Thiagarajar College of Engineering, India) and R. Anand (Nehru Institute of Engineering and Technology, India)
Engineering Science Reference • © 2022 • 490pp • H/C (ISBN: 9781799896364) • US $325.00

Handbook of Research on Challenges and Risks Involved in Deploying 6G and NextGen Networks
A M Viswa Bharathy (Malla Reddy College of Engineering and Technology, India) and Basim Alhadidi (Al-Balqa Applied University, Jordan)
Information Science Reference • © 2022 • 400pp • H/C (ISBN: 9781668438046) • US $275.00

Implementing Data Analytics and Architectures for Next Generation Wireless Communications
Chintan Bhatt (Charotar University of Science and Technology, India) Neeraj Kumar (Thapar University, India) Ali Kashif Bashir (Manchester Metropolitan University, UK) and Mamoun Alazab (Charles Darwin University, Australia)
Information Science Reference • © 2022 • 227pp • H/C (ISBN: 9781799869887) • US $215.00

For an entire list of titles in this series, please visit:
www.igi-global.com/book-series/advances-wireless-technologies-telecommunication/73684

701 East Chocolate Avenue, Hershey, PA 17033, USA
Tel: 717-533-8845 x100 • Fax: 717-533-8661
E-Mail: cust@igi-global.com • www.igi-global.com

To my teachers who taught me how to learn for the sake of learning and my students who taught me how to learn for the sake of teaching.

To my family: with gratitude for your love and support and, my parents for believing in me.

Marcel Ohanga Odhiambo

Table of Contents

Detailed Table of Contents

Chapter 1

    Marcel Ohanga Odhiambo, Mangosuthu University of Technology,
        South Africa
    Weston Mwashita, Vaal University of Technology, South Africa

The internet of things (IoT) revolution is affecting a wide range of academic and industrial disciplines in positive ways. Consumer applications like smart home devices and wearables are giving way to mission-critical applications like public safety, emergency response, industrial automation, self-driving cars, and the internet of medical things. This chapter provides an overview of the internet of things as well as its history. Even though the IoT market is booming, several obstacles are keeping the technology from reaching its full potential. Many of the issues that exist are highlighted and clearly explained in this chapter, with the goal of making it easier for a wide range of scholars/researchers to provide feasible solutions to the challenges. Businesses who embrace IoT ideas and learn to harness the data generated by the internet of things will survive and thrive in the future.

Chapter 2

    Richard B. Watson, Ryan Watson Consulting Pty. Ltd., Australia
    Peter J. Ryan, Ryan Watson Consulting Pty. Ltd., Australia

The internet of things (IoT) is a global ecosystem of networked "things." It is the subject of much research worldwide, although it still has many challenges to overcome before it can achieve its full potential. Many papers have been written on the IoT and related areas including big data analytics, smart cities, and industrial IoT (IIoT). These challenges have mostly been seen as technical, although the IoT's business and societal challenges are also important. Most authors of research papers discuss the research challenges with which they are most familiar, but a framework which identifies and classifies all the challenges and cross-references the publications describing them in detail, is much needed. In this chapter, the authors extend their earlier IoT classification scheme to include more recent papers, and business and societal challenges as well as technical ones. The nature of the classification scheme and research challenges are described; however, the other chapters of this book cover in more detail the individual challenges and proposed strategies to mitigate them.

Chapter 3
Anjum Sheikh Qureshi, Rajiv Gandhi College of Engineering Research and Technology, Chandrapur, India

Internet of things (IoT) is an evolving technology that has interconnected devices and humans across the world. The number of IoT users and devices is growing rapidly. The versatile nature of IoT has led to the development of many new applications. Establishing communication among the interconnected devices at any time from any place is the primary objective of IoT. With the increasing usage of IoT devices, it has become difficult to handle them and monitor their communication process. Some of the challenges faced by the IoT networks are data management, energy consumption, connectivity, security, and addressing of devices. It is essential to overcome these challenges to increase the popularity and acceptance level among the existing users and convince others to adopt IoT into their daily lives.

Chapter 4
Weston Mwashita, Vaal University of Technology, South Africa
Marcel Ohanga Odhiambo, Mangosuthu University of Technology, South Africa

This research presents a power control method for proximity services (ProSe)-enabled sensors to enable the sensors to be seamlessly incorporated into 5G mobile networks. 5G networks are expected to create an enabling environment for 21st-century advancements such as the internet of things (IoT). Sensors are crucial in the internet of things. A power control strategy involving two power control mechanisms is proposed in this research work: an open loop power control (OLPC) mechanism that can be used by a ProSe-enabled sensor to establish communication with a base station (BS) and a closed loop power control (CLPC) mechanism that can be used by a BS to establish the transmit power levels for devices involved in a device to device (D2D) communication. Several studies have proposed power control strategies to mitigate interference in D2D-enabled mobile networks, but none has attempted to address the interference caused by ProSe-enabled sensors communicating with smart phones and 5G BSs.

Chapter 5

In this chapter, the author looks at the challenges to the IoT system due to standard essential patents (SEPs) by looking at guidelines issued by regulators across the world to enable policymakers and judiciaries to deal with critical issues raised in cases involving SEPs. SEPs present a unique challenge as they require balancing the principles of intellectual property law and competition policy. The author analyses four critical challenges raised in disputes involving SEPs by looking at policy guidelines and arrives at the best practices drawn from these guidelines so that they may be used as guideposts for policymakers and regulators to resolve the increasing number of disputes involving SEPs. Finally, the author identifies some key challenges and systemic issues that are yet to be addressed – issues at the centre of some of the most significant disputes involving SEPs today.

Chapter 6

IoT is an amalgamation of diverse devices. The system aims to overcome the infrastructure of the devices. The instruments communicate with each other to accomplish a task. The previous contribution aims in preserving privacy among the communicating devices. It supports formation of cluster, where the device chosen as a cluster head mediates the communication. The user with the devices will be able to post their queries to the untrusted server by camouflaging themselves. The untrusted server responds to the queries which are communicated to the users through the cluster head. Security is vital to the network. The attacks if detected at an early stage can conserve large amount of energy. The current proposal works to enhance reliability to the network by 4.96%, 1.31% enhancement in detecting the attacks, and conserves energy by 6.13% compared to the previous contribution.

Chapter 7

Diverse forms of cyber security techniques are at the forefront of triggering digital security innovations, whereas cybersecurity has become one of the key areas of the internet of things (IoT). The IoT cybersecurity mitigates cybersecurity risk for organizations and users through tools such as blockchain, intelligent logistics, and smart home management. Literature has not provided the main streams of IoT cyber risk management trends, to cross referencing the diverse sectors involved of health, education, business, and energy, for example. This study aims to understanding the interplay between IoT cyber security and those distinct sector issues. It aims at identifying research trends in the field through a systematic bibliometric literature review (LRSB) of research on IoT cyber and security. The results were synthesized across current research subthemes. The results were synthesized across subthemes. The originality of the paper relies on its LRSB method, together with extant review of articles that have not been categorized so far. Implications for future research are suggested.

Chapter 8

Because privacy concerns in IoT devices are the most sensitive of all the difficulties, such an extreme growth in IoT usage has an impact on the privacy and life spans of IoT devices, because until now, all devices communicated one to one, resulting in high traffic that may shorten the life of unit nodes. In addition, delivering data repeatedly increases the likelihood of an attacker attacking the system. Such traffic may exacerbate security concerns. The employment of an aggregator in the system as an intermediary between end nodes and the sink may overcome these problems. In any system with numerous sensors or nodes and a common controller or sink, we can use an intermediate device to combine all of the individual sensor data and deliver it to the sink in a single packet. Aggregator is the name given to such a device or component. Data aggregation is carried out to decrease traffic or communication overhead. In general, this strategy helps to extend the life of a node while also reducing network transmission.

Chapter 9

The technology space has seen the emergence of several buzzwords, including but not limited to artificial intelligence, big data, the internet of things, and robotics. This chapter seeks to discuss the concept of the internet of things and break the very gigantesque details into understandable bits. It examines the concept of the internet of things from the standpoint of a legal practitioner who is practicing in any given jurisdiction and talks about the legal issues that attend the internet of things. Due to the fact that the internet of things refer to the interconnectivity of devices, the legal issue of data privacy and protection is discussed. Other issues like antitrust and 'who-bears-liability' are also equally discussed. A brief insight is equally given along the lines of how these legal issues preclude the smooth rolling out of these technologies in cross-border terms and how industry players are attempting to deal with the issues.

Preface

The Internet of Things (IoT) refers to physical objects (or groups of such objects) that are equipped with sensors, processing power, software, and other technologies and may communicate with other devices and systems over the Internet or other communication networks. Due to the merging of numerous technologies, such as ubiquitous computing, commodity sensors, increasingly sophisticated embedded systems, and machine learning, the area has progressed. The Internet of Things is enabled by traditional domains such as embedded systems, wireless sensor networks, control systems, and automation (including home and building automation). In the consumer market, IoT technology is most closely associated with products that support the concept of the "smart home," such as lighting fixtures, thermostats, home security systems and cameras, and other home appliances that can be controlled by devices associated with that ecosystem, such as smartphones and smart speakers. The Internet of Things can also be used in healthcare. There are several concerns about the risks associated with the growth of IoT technologies and products, particularly in the areas of privacy and security, and as a result, industry and government efforts to address these concerns have begun, including the creation of international and local standards, guidelines, and regulatory frameworks.

The IoT revolution is positively impacting a variety of academic and industrial disciplines. With the world population growing at an alarming rate and being expected to hit 8.5 billion by 2030, it is logical to embed IoT technologies in agricultural activities so that agricultural yields that match the growing population maybe achieved. Environmental parameters can be controlled in smart greenhouses by the utilisation of smart sensors that send information to cloud servers for further processing.

Modern healthcare systems are incorporating IoT technologies to provide a more personalised healthcare system that enables the remote diagnosis and treatment of patients. In a bid to tackle the health challenges brought about by the COVID-19 pandemic, several countries have utilised IoT technologies in the early diagnosis of COVID- 19 cases. IoT technologies are also being used in the monitoring of patients round the clock to keep the disease under check. South Korea has been very successful in this regard.

IoT technologies are also being used in the re-engineering of a variety of products resulting in better performance, reduced cost, and an improved customer experience. Since IoT technologies reduce human-to-computer or human-to-human interactions, mistakes are eliminated. IoT technologies allow equipment to work 24 hours with no need for off-days or rest periods required by people, thereby increasing productivity whilst reducing costs in the process.

With the population of cities across the globe increasing exponentially, the responsible authorities have turned to IoT to convert their cities into smart cities that can intelligently deal with the consequences of overpopulation. IoT technologies allow the establishment of environmentally friendly and energy-efficient infrastructure in smart cities. Traffic flow is optimised by use of IoT technologies that can automatically adjust traffic lights to suit the prevailing conditions. Smart parking aids drivers to park their cars using information from their smart phones. City authorities have also turned to IoT technologies to optimise waste collection efficiency in a bid to reduce costs and address environmental issues.

The objective of this book is to present a survey of mitigating factors to the full realization of the Internet of Things (IoT). The focus is on highlighting the mitigating factors. The contributing authors highlighted some of the factors such as Security, Accessibility, Fault tolerance, Authentication but not necessarily all the mitigating factors to the full realization of the Internet of Things are included in this book. This conceptual book, which is unique in the field of IoT, will assist researchers and professionals working in the area of IoT to better assess IoT and come up with mechanisms to make the full realization of IoT a reality.

The book chapters examine mitigating obstacles to the complete realization of the Internet of Things. This research area has not been sufficiently researched, a small number of scholars have devoted their efforts in this area, most scholars focus solely on IoT applications. Nine book chapters were contributed by the participating authors and are included in this book. The book chapters examine the lack of complete realization of IoT, a topic that few researchers have delved into.

ORGANIZATION OF THE BOOK

The book is organised into nine chapters. A brief description of each of the chapters follows:

Chapter 1 provides an overview of the Internet of Things as well as its history. Even though the IoT market is booming, several obstacles prevent the technology from reaching its full potential. Many of the issues that exist are highlighted and clearly explained in this chapter, with the goal of making it easier for a wide range of scholars/researchers to provide feasible solutions to the challenges. Businesses that embrace IoT ideas and learn to harness the data generated by the Internet of Things will survive and thrive in the future.

Chapter 2 presents a framework which classifies all the challenges that most researchers currently are currently aware of and cross-references key publications describing them in detail. It extends an earlier IoT classification scheme by the authors to include more recent papers, and business and societal challenges as well as technical ones.

Chapter 3 identifies some of the challenges faced by the IoT networks like data management, security, connectivity, energy consumption, and addressing of the IoT devices. The chapter also presents some possible solutions to overcome these challenges. It may not be possible to overcome all the challenges at a time. But to expand the usage of IoT, it is essential to keep on updating oneself with the probable challenges, their causes, and methods to overcome them.

Chapter 4 presents a power control method for Proximity Services (ProSe)-enabled sensors to enable the sensors to be seamlessly incorporated into 5G mobile networks. 5G networks are expected to create an enabling environment for the 21st century advancements such as the Internet of Things (IoT). Sensors are crucial in IoT. A power control strategy involving two power control mechanisms is proposed in this research work: an open loop power control (OLPC) mechanism that can be used by a ProSe-enabled sensor to establish communication with a base station (BS), and a closed loop power control (CLPC) mechanism that can be used by a BS to establish the transmit power levels for devices involved in a device to device (D2D) communication.

In Chapter 5, the author looks at the challenges to the IoT system due to Standard Essential Patents (SEPs) by looking at guidelines issued by regulators across the world to enable policymakers and judiciaries to deal with critical issues raised in cases involving SEPs. SEPs present a unique challenge as they require balancing the principles of intellectual property law and competition policy. The author analyses four critical challenges raised in disputes involving SEPs by looking at policy guidelines and arrives at the best practices drawn from these guidelines so that they may be used as guidelines for policymakers and regulators to resolve the increasing number of disputes involving SEPs.

Chapter 6 presents a security enhancement technique that can be incorporated in an IoT network where the reliability of the network is enhanced by incorporating a hashing concept. The devices suffix the hashed value using an identification, location details, and time of transmission. The hashed code cannot be duplicated as the time and location vary with time. This results in attacks being detected at an early stage.

Chapter 7 evaluates the correlation between IoT cyber security and distinct security issues in various sectors. A Systematic Bibliometric Literature Review (LRSB) was conducted to identify and synthesis data on IoT cyber and security trends.

The findings contribute to a better understanding of the threats and attacks on IoT infrastructure and provide information that can be used to improve cyber defence.

Chapter 8 chapter discusses the differences between IoT privacy and security concerns, as well as several IoT privacy-preserving approaches like as anonymization, dummies, caching (Siddiqui, collaborating, and others. Location, encryption, and homomorphic approaches are all types of anonymization strategies for privacy discussed in this chapter. The authors also went through each layer of the IoT's privacy protection strategy in depth.

Chapter 9 presents a discussion on legal issues surrounding the concept of the Internet of Things. It examines the concept of the Internet of Things from the standpoint of a Legal Practitioner who is practising in any given jurisdiction and talks about the legal issues that attend the Internet of Things. Since the Internet of Things refer to the interconnectivity of devices, the legal issue of Data Privacy and Protection is discussed. Other issues like Antitrust, and 'who-bears-liability' are equally also discussed. A brief insight is equally given along the lines of how these legal issues preclude the smooth rolling out of these technologies in cross-border terms, and how industry players are attempting to deal with the issues.

This book provides an insight into the challenges to the full realization of IoT. The book is dedicated to addressing the major challenges in realizing IoT-based applications including obstacles that vary from cost and energy efficiency to availability to service quality. The aim of the book is to focus on the practical challenges in IoT applications that are enabled and supported by wireless sensor networks and cellular networks. Targeted readers are from varying disciplines who are interested in implementing IoT applications. A number of authors made valuable contribution to the book chapters. The chapters may not necessarily cover all the challenges to the full realization of IoT but a sample as contributed by the respective contributing authors.

- Provides an up-to-date research and applications related to IoT.
- Provides challenges facing IoT scientists and provides ways to solve them in critical daily life issues

We do hope you'll enjoy reading this book and should you have comments, feel free to send them to the publisher/Editor.

Marcel Ohanga Odhiambo
Mangosuthu University of Technology, South Africa

Weston Mwashita
Vaal University of Technology, South Africa

Introduction

This book gives a thorough and high-level examination of the challenges that the Internet of Things (IoT) faces and how they are being addressed. The purpose of the book is to look at and evaluate some of the most important challenges in the development and implementation of real-world IoT applications. Some of the challenges explored in the book's chapters include ensuring and enforcing security and privacy standards, enabling interoperability among multiple disparate protocols and devices, and optimizing the processing power and memory requirements of tiny objects. To effectively manage these issues, the book provides a combination of theoretical and practical research work done by professionals in the field of IoT.

IoT is a significant topic with considerable technical, social, and economic implications. Consumer products, durables, automobiles, industrial and utility components, sensors, and other ordinary objects are all becoming connected to the Internet and have powerful analytical capabilities, influencing how people think, eat, and interact. According to some estimates, the Internet of Things' impact on the Internet and economy will reach 100 billion linked IoT devices by 2025, with a global economic impact of more than $11 trillion.

The possibilities for connecting billions of people and devices to massive amounts of data, storage space, and processing power are endless. These technologies have the potential to improve not only the efficiency of people's professions and many other aspects of their lives, but also the quality of life for people all over the world. These innovations have the potential to lead to a long-term carbon-free world. In a sustainable future, the innovations might be used to track a product's lifecycle, aid in the supply of crucial equipment and medical supplies in dangerous environments such as warzones, and even foresee and avert natural disasters. At the same time, considerable barriers stand in the way of IoT fulfilling its full potential.

Reports of Internet-connected device hacking, surveillance issues, and privacy concerns have already attracted the public's interest. Policy changes, legal challenges, and development issues have all emerged in addition to technological issues. This book provides a guide to readers of the Internet Society through the discussion over the Internet of Things, particularly considering contrasting forecasts regarding its benefits and risks. The Internet of Things encompasses a wide range of concepts that are complicated and interconnected from several angles.

The IoT has recently seen a real concern with the proliferation of monitoring systems such as smart surroundings, smart automobiles, and smart wearable gadgets. People's lives have been transformed which has made them more adaptive and complex. Intelligent technology, for example, will improve the performance of doctors, nurses, patients, and the healthcare business in a healthcare monitoring system. The IoT revolution, often known as the fourth industrial revolution, is expected to transform how humans interact with technology and pave the way for high-tech machine-to-machine communication

Before IoT technologies can be broadly used in the real world, several challenges and concerns must be overcome. These issues are technological as well as societal in character. Assuring interoperability among varied networked things, giving devices with a high level of smartness through autonomous and flexible computing, and maintaining user confidence, security, and privacy are among the most pressing concerns. It's extremely important in the IoT to make optimum use of processing power and memory space in small and resource-constrained devices and objects.

Buildings, both residential and commercial, are undergoing significant transformations, and IoT technologies are having an impact on their future. In a range of applications and contexts, researchers have lately used IoT technology to turn traditional buildings into smart, efficient, and secure structures. Despite the advancement of successful IoT technologies, IoT applications and procedures must still be improved to realize the full potential of the technology. This book aims to fill in the gaps in the current literature and serve as a foundation for future research on the topic.

In a few contexts, the Internet of Things will connect people and machines, allowing important insights to be obtained from the collection and analysis of user data. Industry 4.0 has introduced a host of new solutions to industry, business, and everyday activities that were previously restricted by cost, energy, and data storage constraints. Given the rapid and often unregulated evolution of cyber technology, businesses should consider developing intrinsically secure systems capable of rigorous authentication and data security to inspire trust and alleviate end users' privacy worries.

Embedding security into a system, lowering costs in industry applications, increasing production speed, and increasing redundancy and flexibility in a system through decentralized data storage systems are just a few of the benefits of incorporating IoT into the fourth industrial revolution. The Internet of Things will enable broad health care, particularly for the elderly, by utilizing smart devices that are no longer constrained by restricted processing and storage capabilities and a finite energy budget. Users also benefit from systems that are built to allow for more user mobility.

Because the Internet of Things is still in its infant stages and the technology has not yet fully matured to support a higher level of connectedness, as promoted by tech enthusiasts and the general media, it's critical to set realistic targets for deploying effective IoT solutions. While there are numerous obstacles to building a viable IoT solution, rapid technological breakthroughs have the potential to accelerate IoT development.

Chapter 1
The Challenges Brought About by the IoT Revolution

Marcel Ohanga Odhiambo
Mangosuthu University of Technology, South Africa

Weston Mwashita
Vaal University of Technology, South Africa

ABSTRACT

The internet of things (IoT) revolution is affecting a wide range of academic and industrial disciplines in positive ways. Consumer applications like smart home devices and wearables are giving way to mission-critical applications like public safety, emergency response, industrial automation, self-driving cars, and the internet of medical things. This chapter provides an overview of the internet of things as well as its history. Even though the IoT market is booming, several obstacles are keeping the technology from reaching its full potential. Many of the issues that exist are highlighted and clearly explained in this chapter, with the goal of making it easier for a wide range of scholars/researchers to provide feasible solutions to the challenges. Businesses who embrace IoT ideas and learn to harness the data generated by the internet of things will survive and thrive in the future.

INTRODUCTION

The IoT revolution is having a favourable impact on a wide range of academic and industrial fields. With the world's population growing at an alarming rate, an estimated 8.5 billion people by 2030, it makes sense to integrate IoT technologies into agricultural activities to increase agricultural output to maintain pace with the

DOI: 10.4018/978-1-7998-9312-7.ch001

rising population amongst other needs. Smart sensors that communicate data to cloud servers for further processing can be used to manage environmental factors in smart greenhouses.

The IoT technology has been implemented into modern healthcare systems to provide more customised healthcare systems that allow for remote diagnosis and treatment of patients. In a bid to tackle the health challenges brought about by COVID-19 pandemic, several countries have utilised IoT technologies in the early diagnosis of COVID-19 cases. IoT technologies are being used in the continuous monitoring of patients to keep the disease under control. South Korea has been very successful in this regard.

IoT technologies are being used in the re-engineering of a variety of products resulting in better performances, reduced costs and improved customer experiences/satisfactions. Since IoT technologies reduce human-to-computer or human-to-human interactions, mistakes can be minimised. IoT technologies enable equipment to continuously work 24 hours without the need for OFF-days as required by humans, thereby increasing productivity whilst reducing costs in the process.

With the population of cities across the globe increasing exponentially, responsible authorities are exploring IoT technologies to convert their cities into smart cities that can intelligently deal with the consequences of overpopulation. IoT technologies enable the establishment of environmentally-friendly and energy-efficient infrastructure in smart cities. Traffic flow is optimised by use of IoT technologies that can automatically adjust traffic lights to suit the prevailing traffic flow conditions. Smart parking systems help drivers to park their cars using information from their smart phones. City authorities have also turned to IoT technologies to optimise waste collection efficiency in a bid to reduce costs and address environmental issues.

IoT implementation is expanding beyond consumer applications like smart home gadgets, wearables to mission-critical applications like public safety, emergency response, industrial automation, self-drive/autonomous cars, and the Internet of Medical Things. Engineers and designers must address application concerns and trade-offs of IoT systems from the design phase, manufacturing and application of IoT systems as IoT systems become more common in mission-critical areas.

BACKGROUND

According to Ranger (2020), IoT refers to the billions of physical devices connected to the Internet, collecting and exchanging data around the world. It is now feasible to turn everything, from a pill to a jet, into a part of the Internet of Things, thanks to the advent of super-cheap computer chips and the widespread availability of wireless networks. Connecting these diverse systems and attaching sensors to them gives the devices that would otherwise be dumb, a level of digital intelligence, enabling them to convey real-time data without the involvement of human beings. The IoT technology is merging the digital and physical worlds together to make the world around us smarter and more responsive. The IoT technology is enabled by traditional domains like embedded systems, wireless sensor networks, control systems, automation (including home and building automation), and others. In the consumer market, IoT technology is most synonymous with products pertaining to the concept of the "smart home", including devices and appliances (such as lighting fixtures, thermostats, home security systems and cameras, and other home appliances) that support one or more common ecosystems, and can be controlled via devices associated with that ecosystem, such as smartphones and smart speakers. There are serious concerns about the dangers in the growth and application of IoT, especially in the areas of privacy and security and, consequently industry and governmental efforts in addressing these concerns are welcome in addition to including the development of international standards.

A Brief History of IoT and Explanation of Important Terms

The notion of the IoT technology was not officially named until 1999. In the early 1980s, a Coca-Cola machine at Carnegie Melon University was one of the first examples of the Internet of Things. Before making the trip, local programmers would connect to the refrigerator through the Internet and check to see whether there was a drink available and if it were cold. By 2013, the Internet of Things had evolved into a system that utilized a variety of technologies, including the Internet, wireless communication, and micro-electromechanical systems (MEMS). The IoT is supported by traditional sectors such as automation (including building and home automation), wireless sensor networks, GPS, control systems, and others.

Kevin Ashton of Procter & Gamble, later MIT's Auto-ID Centre, created the term "Internet of Things" in 1999. He envisioned radio-frequency identification (RFID) as critical to the Internet of Devices at the time, as it would allow computers to manage all the individual things. The Internet of Things' fundamental idea is to implant short-range mobile transceivers in a variety of gadgets and everyday requirements to enable new types of communication between people and things, as well as between things themselves. Defining the Internet of Things as "simply the point in time when more 'things or objects' were connected to the Internet than people", Cisco Systems estimated that the IoT was "born" between 2008 and 2009, with the things/people ratio growing from 0.08 in 2003 to 1.84 in 2010.

Accessibility

In the context of IoT, accessibility refers to the design of products, gadgets, services, or places such that they are usable by individuals. The "ability to access" and benefit from a system or entity is referred to as accessibility. The notion focuses on facilitating access for those with disabilities or special needs or facilitating access using assistive technology; nonetheless, accessibility research and development benefits everyone.

Fault-tolerance

Fault tolerance is the ability of a system to continue to function properly even if some of its components fail (or have one or more faults within them). When compared to a naively constructed system, where even a minor failure could cause entire breakdown, the drop in operating quality is proportional to the severity of the failure. In high-availability or life-critical systems, fault tolerance is especially important. Graceful degradation refers to a system's capacity to sustain functionality even when other parts of it fail.

When an element of a system fails, a fault-tolerant design allows the system to continue operating, albeit at a lower level or downgraded, rather than failing totally. In the event of a partial failure, the term is most typically used to describe computer systems built to remain completely operational with, maybe, a drop in throughput or an increase in response time. That is, the system does not come to a halt owing to hardware or software issues. A motor vehicle built to remain drivable even if one of its tyres is punctured, or a structure capable of maintaining its integrity in the face of damage caused by fatigue, corrosion, manufacturing faults, or impact, are examples from another fields.

Fault tolerance can be achieved within the limits of an individual system by predicting exceptional conditions and designing methods to deal with them, as well as aiming for self-stabilisation such that the system converges towards an error-free state. If the implications of a system failure are catastrophic, or the expense of making it sufficiently dependable is prohibitively great, using some form of duplication may be a preferable alternative. In any case, if the consequences of a system failure are so severe, the system must be able to revert to a safe mode by reverse process. If humans are included in the loop, this is similar to roll-back recovery, but it can entirely be a human action.

Authentication

The act of proving an assertion, such as the identity of a computer system user, is known as authentication (from Greek: authentikos, "actual, genuine," from authentes, "author"). Authentication is the process of verifying a person's or thing's identity, as opposed to identification, which is the act of indicating that person's or thing's identity. Validating personal identity documents, establishing the validity of a website with a digital certificate, carbon dating an artifact to determine its age, or ensuring that a product or document is not counterfeit are just a few examples. The required authentication is conducted on each piece. Authentication definition is the act or process of certifying something as genuine or authoritative.

Privacy

An individual's or a group's ability to seclude themselves or information about themselves, and therefore express themselves selectively, is known as privacy. When a person refers to something as "private," it usually refers to anything that is unique or sensitive to them. The concept of responsible use and protection of information falls under the area of privacy, which is partially overlapped with security. Many countries' privacy laws, and in some cases, constitutions, include the right not to be exposed to unjustified intrusions of private access by the government, corporations, or individuals. In the business world, a person may provide personal information, such as for advertising, in exchange for a reward, while an individual may be subject to public interest rules. Identity theft can occur when personal information is voluntarily supplied but then stolen or exploited. Universal individual privacy is a new concept linked primarily with Western society, particularly British and North American culture, and has remained completely unknown in some societies until recently. Most cultures, on the other hand, acknowledge an individual's ability to keep certain aspects of their personal information hidden from the rest of society, which can be likened to closing the door to their home.

Data Integrity

According to Talend, Data integrity is a vital part of the design, implementation, and use of any system that stores, processes, and/or retrieves data because it ensures data accuracy and consistency throughout its life cycle. The phrase has a broad scope and can have a variety of connotations depending on the context - even when used in the same context as in computing. While data validation is a pre-requisite for data integrity, it is sometimes used as a surrogate term for data quality. The opposite of data corruption is data integrity. Any data integrity technique has the same goal:

to verify that data is recorded exactly as intended (such as a database correctly rejecting mutually exclusive possibilities). Furthermore, guarantee that the data is the same as it was when it was originally recorded and when it is retrieved later. In a nutshell, data integrity seeks to prevent unintended data alterations. Data integrity should not be confused with data security, which is the practice of safeguarding data against unwanted access. Data integrity failure is defined as any unwanted changes to data due to storage, retrieval, or processing action, including malicious intent, unanticipated hardware failure, and human mistake. If the modifications are the consequence of unauthorized access, data security may have failed. Depending on the data involved, this could range from a minor change in the colour of a single pixel in an image to the loss of vacation photos or a business-critical database, and even the catastrophic loss of human life in a life-critical system.

Quality of Service (QoS)

Users of telecommunication services expect network service providers to provide reasonable and acceptable levels of quality of service to ensure that their communication is not degraded (delayed, lost, refused access, etc.) in the network. As a result, service providers must create techniques to ensure network quality of QoS and end-to-end delivery of users' messages throughout the network by ensuring network resources match the required services along the data path. Recent technology advancements in the telecommunications area are fast converging towards integrated voice, data, and video traffic, ATM and IP coexistence, and wired and wireless service integration, with the ultimate goal of delivering new and advanced broadband services to users (Aurelio *et al.* 2001). This is mostly due to the emergence of multimedia applications, which face new and complex problems in terms of QoS prediction, assurance, and adaptation. QoS can be defined in a variety of ways and encompasses a wide range of service criteria such as performance, availability, dependability, security, bandwidth, congestion, routing, stability, delays, and pricing (Firoiu *et al.,* 2002). Service management quality is a crucial part of performance management. QoS management applications track performance within the network as well as at the "edge" access point where consumer services are delivered, to detect performance degradation. Network QoS management necessitates the availability of management information to determine network load under both natural and artificial conditions, as well as the gathering of performance data to provide statistics and facilitate capacity planning. Performance management requires access to a significant amount of network status data with minimal disruption to the managed network. As a result of the status information, elements of the network may be reconfigured to relieve congestion e.g., by changing the routing strategy, pricing strategy, re-allocating resources, such as bandwidth, etc. Guaranteed QoS is required for the implementation of effective

Figure 1. Application areas of IoT technologies
(Nižetić et al. 2020)

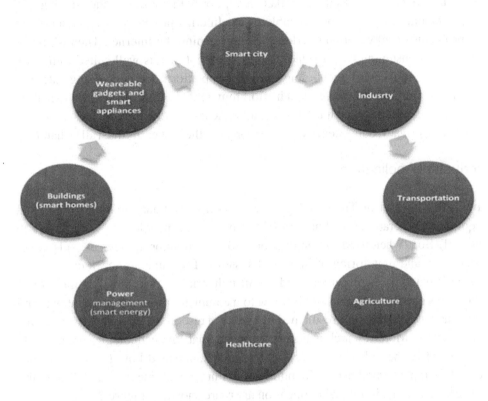

IoT services. IoT devices connect to the Internet and share data and information to support IoT services. The implementation of effective IoT services necessitates adherence to quality of service.

Applications of IoT

The diverse areas of applications for IoT devices are often divided into consumer, commercial, industrial, and infrastructure spaces as shown in Figure 1.

Consumer Applications

Connected automobiles, home automation, wearable technology, connected health, and appliances with remote monitoring capabilities are among the many IoT products being developed for consumer usage. (Odhiambo *et al.* 2012) point out that a growing segment of the Internet is electronic commerce. Consumers are looking for online vendors who sell items and services. Meanwhile, suppliers are on the lookout for new customers to expand their market share. For both consumer and supplier, the large amount of information available on the Internet produces a tremendous deal of problems or information overload when searching the Internet. The task is not only time consuming but also tedious. This type of task is well-suited to the IoT infrastructure. The IoT technology provides a platform for conducting Internet-based E-Commerce. However, much effort remains to be done to make this a reality. Security, accessibility, fault tolerance, authentication, infrastructure, costs for used resources, privacy, data integrity, and a variety of other concerns must all be handled.

Industrial Applications

Industrial Internet of Things (IIoT) devices collect and analyse data from linked equipment, operational technology (OT), places, and people, and are referred to as IIoT. IIoT, when used in conjunction with OT monitoring devices, aids in the regulation and monitoring of industrial systems. The same approach can be used for automated asset placement updates in industrial storage units, as the size of the assets can range from a small screw to a complete motor replacement part and misplacing such assets can result in a huge waste of resources (personnel efforts, time, and money). More research is needed to ensure that IoT technologies are properly integrated in the industry, as well as to better understand how IoT technologies could be implemented in certain industries to improve productivity and availability of services. Some industrial application areas are shown in Figure 2.

Infrastructure Applications

An important use of IoT is monitoring and regulating operations of sustainable urban and rural infrastructures such as bridges, railway tracks, and on and offshore wind farms. The IoT infrastructure can be used to track occurring events or changes in structural conditions that could jeopardise safety and put people's lives in danger and/or compromise the safety of the systems. The construction industry can benefit from the IoT technology in terms of cost savings, time savings, better quality workdays, paperless workflows, and increased productivity. With teal-time data analytics, one can make faster judgments and save money. It can also be used to efficiently schedule repair and maintenance activities by coordinating duties amongst multiple service providers and facility users. IoT devices can also be used to regulate essential infrastructure, such as bridges, which allow ships to pass through. In all infrastructure-related fields, the use of IoT devices for monitoring and running infrastructure is anticipated to improve incident management and emergency response coordination, as well as quality of service, up-times, and cost of operation.

Even garbage management can benefit from the automation and optimization that the Internet of Things can bring.

Figure 2. General concept of IoT industrial application
(Aazam et al., 2018)

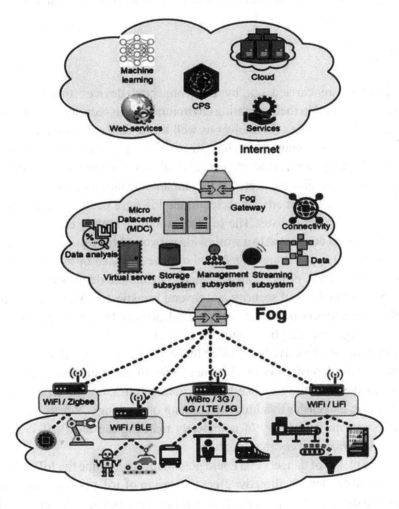

FACTORS IMPACTING THE IMPLEMENTATION OF IOT

This section captures some but not all factors that impact the implementation of IoT. The challenges with IoT go beyond making and connecting devices that work (Arpan & Balamuralidhar, 2017). The integrated product and services need to work seamlessly, almost invisible to the end user. As Weiser (1991), suggests in his ubiquity paradigm, that people need machines that fit the human environment instead of forcing humans to enter theirs. The service must meet user needs, integrate effortlessly into daily life or industrial processes, and improve the user's life or business process. This necessitates the dependability and robustness of all integrated system components. Occasionally, the system is over-engineered to meet the application's objectives. The design of the system should be trimmed and adjusted to the business goals and deployment scenario.

Security

IoT operations are carried out by interconnected devices that communicate information acquired in their operating environment. The security concerns involve the security of the devices themselves as well as the data or information collected by IoT devices and communication infrastructures. When security is breached, IoT technology becomes a threat to human survival, since a breach of IoT devices or activities could have disastrous effects for humans and the environment (Mohamed *et al.,* 2020). Devices that are connected to the IoT have become an indispensable element in people's daily lives. The IoT technology is rapidly expanding as more and more items get connected to a worldwide network. Many IoT devices have very sensitive data and applications that should only be accessed by authorised users

The security of IoT devices should extend beyond the devices themselves. IoT devices have a low level of security and several weaknesses. Many people believe that IoT manufacturers don't put security and privacy first. Despite the security concerns, though, IoT adoption continues to grow.

The current IoT objective is to ensure that everything, from wherever, is always connected to the Internet via the Internet Protocol (Samaila *et al.,* 2018). This concept has the potential to make houses, communities, and electric grids safer, more efficient, and easier to administer, among other things. Nonetheless, several roadblocks remain in the way of completely realising the ideal IoT. Security of systems and devices is very important.

There are plenty of dangers that could potentially undermine the IoT technology. Some of these challenges, directly affect the design of IoT systems. Spoofing, for example, is a threat that happens when an attacker compromises a lower-level device with little or no protection and gains access to a network with secured equipment, which is then duped into thinking the invader is encrypted. To attack IoT devices, attackers use valid Address Resolution Protocol packets, which traditional detection systems may find difficult to detect. Attackers can use spoofing to link numerous IP addresses to a single Media Access Control address (MAC). As a result, the targeted MAC address will be flooded with traffic meant for alternative IP addresses as shown

Figure 3. Spoofing attack
(Husain et al., 2020)

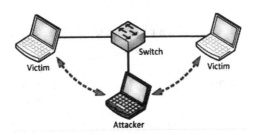

in Figure 3. There is a necessity for detection systems that employ non-traditional methods for detecting spoofing threats in IoT networks.

The most prevalent and easiest-to-implement assaults on IoT systems are denial-of-service (DoS) attacks. They come in a variety of shapes and sizes and are described as any attack that threatens the network's or systems' ability to perform as planned. Majority of IoT devices do not receive the same software protection updates that connected PCs do on a regular basis. As a result, as demonstrated in Figure 4, IoT are quickly changed into infected zombies and used as weapons to send huge amounts of data.

According to Husain *et al.* (2020), to mitigate DoS attacks, some networks make use of the (Border Gateway Protocol (BGP) and Flow Specification (Flow spec) Route Reflector features.

Challenges in Sensing

Sensing is a critical component of IoT and wireless sensor networks. In a typical scenario, sensed data is delivered via an IoT network for post-processing and inference to gain insights. The precision of sensors is critical for post-processing inference to be useful. International Organization for Standardisation standard 5725:1994 distinguishes between precision and trueness in this regard, while highlighting the sensor's integrity (Suzuki *et al.*, 2019). Radio Frequency (RF) sensing, on the other hand, uses channel state information for sensing and, because of its inherent electromagnetic nature, relies on machine learning to classify sensed data, posing additional challenges like linearity, repeatability, resolution, hysteresis, temperature coefficients, stability, and calibration when addressing these issues, the dependability of RF sensing improves, paving the way for a possible future for ambient RF sensing. RF sensing also has the advantage of being haptic, making it suitable for a wide range of applications, as well as being simple to implement. Following developments in machine learning algorithms and RF characterisation, RF

Figure 4. Architecture of a DoS attack
(Husain et al. 2020)

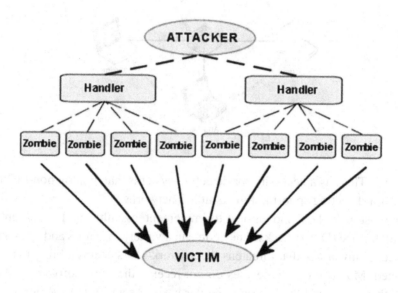

sensing is expected to become a fundamental feature of IoT networks. This can be seen in recent improvements in the THz spectrum for sensing and communications.

Limited Processing Power

The majority of IoT apps consume a small amount of data. This saves money and extends battery life, but it makes it impossible to update over-the-air and it prohibits the device from employing security tools like firewalls and virus scanners. As a result, they are more prone to hacking. Low-power apps that were not designed to connect to the cloud are likely to be vulnerable to modern cyber threats. These outdated materials, for example, might not be compatible with contemporary encryption standards. Making obsolete apps Internet-ready without making significant changes is dangerous. Each IoT application has unique requirements for the range within which it is expected to send sensed data. This, too, is heavily influenced by the geographic limits of a given application. To transmit data over longer distances, more power must be injected into radio communication, resulting in higher battery use. As a result, to reduce power consumption, the application scenario must be expressly specified using a green methodology. A device's energy consumption is also determined by how long it is continuously sensing, processing, and transmitting/receiving potential data. More sensor samples result in improved data interpretation, but they also increase power consumption.

Malicious Insider Attack

It is possible to obtain users' personal information from within an IoT ecosystem. It is done by an authorised user to gain access to another user's information. It is a unique and complicated attack that necessitates a variety of defence systems (Sanzgiri & Dasgupta, 2016). Even with improved security threat detection systems, insider threats might be difficult to detect. This is most likely because an insider danger rarely manifests itself until after the attack. Furthermore, because the malicious actor seems to be a legitimate user, distinguishing between typical and suspicious behaviour in the days, weeks, and months leading up to an attack can be challenging. Because insiders have authenticated access to sensitive data, the insider exploit may not be discovered until the data has vanished. Continuous monitoring tools and attack surface management aid in the prevention of an insider attack by scanning computing systems and networks on a regular basis, taking inventory of vulnerabilities, prioritizing them, and alerting users when action is required.

Standardisation

Most current Internet standards lacked the foresight to include the IoT technology, which is a relatively new notion. Until now, a few significant firms have been responsible for the definition of the IoT protocols or have undertaken the requisite research and development, and the market presence to enforce those standards. Standards are necessary for different devices to communicate with one another. Regulatory standards are essential if information is privileged and confidential to ensure that it remains secure and to specify who owns it and under what conditions it can be shared to others. There are quite several other reasons why the IoT regulations are important.

Security and Reliability

According to Lionel (2018), about 50 billion IoT devices will be in use around the world by 2030, resulting in a huge web of interconnected gadgets ranging from smartphones to household appliances. The majority of the world's over 25 billion IoT linked devices are not genuinely secure. When security is lacking in even seemingly innocuous household appliances or other IoT items, chronic vulnerabilities and hazards arise. When connecting individual devices to a network, the entire system is only as secure as its weakest component. As a result, a low-cost product with a major security flaw might endanger the entire network. This is one of the reasons why IoT consumers are calling for standardisation.

The Complexity of Devices

Without global standards, the complexity of devices that need to connect and communicate with one another (together with all the addressing, automation, quality of service, interfaces, data repository and associated directory services, and so on) will skyrocket. To function at a degree of complexity that is acceptable, manageable, and scalable, common standards are essential.

Interoperability

The ability of devices to communicate with one another is a fundamental requirement of the IoT technology. This is aided by standard-compliant products, technology, and services. IoT is made up of a wide range of devices far more than the traditional Internet. To support multi-vendor solutions and the integration of heterogeneous devices, standardised communication protocols can be programmed on diverse commodity, off-the-shelf hardware (i.e. chipsets, gateways). In addition to fostering long-term interoperability, this helps end users avoid the business risks of vendor lock-in, in which a single provider maintains complete control over functionality design and future product/technology advancement. Two levels of interoperability are required for smooth operation in IoT networks: device interoperability and network interoperability. Device interoperability refers to the capacity of heterogeneous IoT devices to integrate and communicate with another using multiple communication protocols and standards. It is concerned with the information transmission across heterogeneous devices and communication protocols, as well as the capacity to add new devices into any IoT platform.

IoT devices, unlike desktop computers, rely on a variety of short-range wireless communication and networking technologies, which are more inconsistent and unpredictable. Interoperability at the network level refers to the procedures that allow for seamless message exchange across systems via many networks (networks of networks) for end-to-end communication. Each system should be able to exchange messages with other systems via various forms of networks to make them interoperable. According to Bello *et al.*, (2016), the network interoperability level should manage challenges such as addressing, routing, resource optimisation, security, QoS, and mobility support due to the dynamic and heterogeneous network environment in IoT. Improving IoT interoperability is critical to the IoT's success. Interoperability between devices and brands is not only a driver for the manufacturers to scale up their product offerings thanks to more cost-effective sourcing of widely adopted off-the-shelf technologies, but it is also a driver for the manufacturers to develop attractive product ecosystems for end users.

Global Scalability

Scalability is essential for meeting the IoT's changing technological and business requirements. All architectures and solutions must be able to deploy for medium-sized instances and then scale up or down effortlessly as needed. This includes network, storage, analytics, and security. Industrial users with global operations seek to develop IoT connection that can be used across all their locations. When an IoT deployment comprises of thousands of devices in various countries, dealing with a variety of connectivity providers and technologies, manually managing each connection is practically impossible. Standardised solutions are generally applicable and help to reduce installation complexity, protecting long-term investment.

Standardisation Makes Life Easier for All

Many gadgets today are not "plug and play" ready, which is one of the repercussions of an unstandardised IoT. To make their existing technology operate, users must download software and drivers. IoT must be made easier to use if many people are to embrace the technology. The simplest approach to do so is to develop open source technologies and Application Programming Interfaces (APIs) that IoT device creators may use to connect their devices to the billions of others on the planet. If these items adopt standard protocols, they will be able to communicate in a way that everyone understands, making life easier for both customers and businesses. Even though many businesses avoid partnering in the name of healthy competition, it is widely acknowledged that a product is only as smart as the people who create it. With so much power at risk, ensuring that all brains are working together to construct the IoT in a safe and purposeful way would benefit the entire planet. Without standards, there is a considerably higher risk that technologies and solutions will fail to perform as planned, will have shorter lifespans, will be incompatible with other solutions and hence be siloed, and will limit IoT technology users to a single vendor with its own proprietary standard.

IoT Standardisation Bodies

According to Arpan *et al.,* (2018), various Standard Development Organizations (SDOs) throughout the world are working to develop standard platforms, protocols, and technologies to ensure that IoT devices work together seamlessly. Different SDOs can be broadly classified into two classes in terms of technology offerings: generic and application specific.

Generic

SDOs like the International Telecommunication Union, Institute of Electrical and Electronics Engineers, Internet Engineering Task Force (IETF), the 3rd Generation Partnership Project and one Machine-to-Machine (oneM2M) have long played a key role in creating technical standards that cover the entire problem space. They have either stated policies or generic reference architectures, or they've proposed a common protocol for communicating. The SDOs also define technology domains. These SDOs are generally open in the sense that anyone can look at the specifications without having to be a member of the organization. To donate, however, one must be a member. The IETF is an exception. It is, in fact, open in the real meaning of the word. Any individual can theoretically contribute to IETF standards, and their contributions are recognized on a meritocratic basis.

Application Specific

For application specific standards, SDOs or alliances have been formed with the goal of standardising technologies for a certain application domain. To develop the communication model, these SDOs rely on existing architectures and protocol options, as well as a general approach. They provide specific standards for various exchange types to fill up common gaps in standard offerings. Within their member organisations, these SDOs are usually closed. Fairhair Alliance is a good example.

Table 1 lists the standardisation bodies that have played a key role in creating technical standards for the IoT issue space. They have either outlined policies or generic reference architectures, or they have proposed a standard protocol for IoT connectivity.

CONCLUSION

This chapter has provided an overview of the IoT technology as well presenting its history. The obstacles that stand in the way of IoT preventing the technology from reaching its full potential have been presented. This was done with an aim of making it easier for a wide range of scholars to give feasible solutions to these challenges. This chapter has highlighted the challenges or impediments to the full implementation of IoT.

Table 1. Standardisation bodies involved in IoT standards

Standardisation Body	Their Roles in IoT Standardisation
Internet Research Task Force (IRTF)	Even though IoT-focused IETF working groups have already developed the first wave of mature IoT standards, new research problems based on their utilization are developing. The IRTF Thing-to-Thing Research Group (T2TRG) was established in 2015 to examine open research concerns on the Internet of Things, with an emphasis on issues with IETF standardization potential.
World Wide Web Consortium (W3C)	The W3C Web of Things continues its promise of establishing a Web-based abstraction layer for existing platforms, devices, gateways, and services in order to combat fragmentation on the Internet of Things. It improves interoperability by complementing current standards, lowering risk for investors and customers.
International Telecommunication Union (ITU-T)	IoT and its applications, including smart cities and communities," is the responsibility of ITU Study Group 20 (SG20). It's also responsible for semantics aspects; big data aspects; detailed requirements of networks supporting IoT applications; accounting and charging aspects; identification, security, and privacy; openness; etc.
IoT Acceleration Consortium	The IoT Acceleration Consortium aims to bring together the strengths of Japanese government, industry, and academia to provide a framework for developing and proving IoT-related innovations, as well as generating and facilitating new business models.
European Committee for Standardization (CEN/CENELEC)	CEN/CENELEC has identified a number of domains (and subdomains) as critical to the Digital Single Market (such as 5G, Internet of Things, Cybersecurity, Cloud, data-driven services and applications), as well as essential interoperability standards in areas like eHealth, Intelligent Transportation Systems, Smart Cities, and Smart and Efficient Energy Use)
European Telecommunications Standards Institute (ETSI)	The European Standards Organization (ESO) is the acknowledged regional standards body for telecommunications, broadcasting, and other electronic communications networks and services. Special interest organizations include: ETSI Smart Cities is part of the ETSI ISG CIM group.
3rd Generation Partnership Project (3GPP)	3GPP completed in June of 2016, the standardization of NB-IOT, the new narrowband radio technology developed for the Internet-of-Things (IoT).
VDI, VDE, IT	VDI/VDE Innovation Technology GmbH is a major provider of IoT innovation and technology-related services.
Alliance for Internet of Things Innovation (AIOTI)	AIOTI was founded in 2016 with the goal of contributing to the development of a dynamic European IoT ecosystem and accelerating IoT adoption. Key European IoT actors, such as large corporations, successful SMEs, and burgeoning start-ups, as well as research institutes, universities, trade organisations, and end-user representatives, are among our members.
Big Data Value Association (BDVA)	The Big Data Value Association is an industry-driven international non-profit organization with over 200 members from across Europe and a well-balanced mix of large, small, and medium-sized businesses, as well as research and user organizations.
Open Platforms Communication Foundation (OPC)	OPC is an interoperability standard for transferring data securely and reliably in the industrial automation and other industries. It is platform agnostic and ensures a smooth flow of data across devices from various manufacturers. The OPC Foundation is in charge of the standard's creation and maintenance.
oneM2M	oneM2M is a global standards initiative for Machine-to-Machine (M2M) and Internet-of-Things (IoT) technologies that encompasses requirements, architecture, API specifications, security solutions, and interoperability. oneM2M was founded in 2012 by eight of the world's most prestigious ICT standards development organizations: ARIB (Japan), ATIS (North America), CCSA (China), ETSI (Europe), TIA (North America), TSDSI (India), TTA (Korea), and TTC (Japan), together with seven industry fora, consortia, or standards bodies (Broadband Forum, CEN, CENELEC, Global Platform, HGI, Next Generation M2M Consortium, OMA) and over 200 member organizations.

REFERENCES

Aazam, M., Zeadally, S., & Harras, K. A. (2018). Deploying fog computing in industrial internet of things and industry 4.0. *IEEE Transactions on Industrial Informatics, 14*(99), 1. doi:10.1109/TII.2018.2855198

Arpan, P., & Balamuralidhar, P. (2017). *Key Factors for a Realistic Internet of Things (IoT)*. https://www.methodsandtools.com/archive/realiot.php

Arpan, P., Hemant, K. R., Samar, S., & Abhijan, B. (2018). *IoT Standardization: The Road Ahead*. https://www.intechopen.com/books/internet-of-things-technology-applications-and-standardization/iot-standardization-the-road-ahead

Aurelio, L. C., Antonio, P., & Orazio, T. (2001). QoS management in programmable networks through mobile agents. *Microprocessors and Microsystems, 25*(2), 111-120. https://www.sciencedirect.com/science/article/pii/S0141933101001041 doi:10.1016/S0141-9331(01)00104-1

Bello, O., Zeadally, S., & Badra, M. (2016). Network layer inter-operation of Device-to-Device communication technologies in Internet of Things (IoT). *Ad Hoc Networks, 0*, 1–11.

Chen, Y.-Q., Zhou, B., Zhang, M., & Chen, C.-M. (2020). Using IoT technology for computer-integrated manufacturing systems in the semiconductor industry. *Applied Soft Computing, 89*, 89. doi:10.1016/j.asoc.2020.106065

Firoiu, V., Le Boudec, J., Towsley, D., & Zhi-Li, Z. (2002). Theories and models for Internet quality of service. *Proceedings of the IEEE, 90*(9), 1565-1591. 10.1109/JPROC.2002.802002

Husain, A., Hamed, A. R., & Wasan, A. (2020). ARP Spoofing Detection for IoT Networks Using Neural Networks. *Proceedings of the Industrial Revolution & Business Management: 11th Annual PwR Doctoral Symposium (PWRDS)*, 1-9.

Lionel, S. V. (n.d.). *Number of internet of things (IoT) connected devices worldwide in 2018, 2025 and 2030*. https://www.statista.com/statistics/802690/worldwide-connected-devices-by-access-technology/

Mohamed, L., Nabil, K., Khalid, M., Abdellah, E., & Mohamed, F. (2020). IoT security: Challenges and countermeasures. *Procedia Computer Science, 177*, 503-508. https://www.sciencedirect.com/science/article/pii/S1877050920323395 doi:10.1016/j.procs.2020.10.069

Nižetić, S., Šolić, P., López-de-Ipiña González-de-Artaza, D., & Patrono, L. (2020). Internet of Things (IoT): Opportunities, issues, and challenges towards a smart and sustainable future. *Journal of Cleaner Production, 274,* 122877. doi:10.1016/j. jclepro.2020.122877 PMID:32834567

Odhiambo Marcel, O., & Umenne, P. O. (2012). Net-Computer: Internet Computer Architecture and its Application in E-Commerce. *Journal: Engineering, Technology & Applied Science Research, 2*(6), 302 – 309.

Ranger, S. (2020). *What is the IoT? Everything you need to know about the Internet of Things right now.* https://www.zdnet.com/article/what-is-the-internet-of-things-everything-you-need-to-know-about-the-iot-right-now/

Samaila, M. G., Neto, M., Fernandes, D. A. B., Freire, M. M., & Inácio, P. R. M. (2018). Challenges of securing Internet of Things devices: A survey. *Security and Privacy, 1*(2), e20. doi:10.1002py2.20

Sanzgiri, A., & Dasgupta, D. Classification of insider threat detection techniques. In *Proceedings of the 11th Annual Cyber and Information Security Research Conference.* ACM.

Suzuki, T., Takeshita, J. I., Ogawa, M., Lu, X.-N., & Ojima, Y. (2020). *Analysis of measurement precision experiment with categorical variables.* Paper presented at 13th International Workshop on Intelligent Statistical Quality Control 2019, IWISQC 2019, Hong Kong, Hong Kong.

Talend. (n.d.). *What is Data Integrity and Why Is It Important?* https://www.talend.com/resources/what-is-data-integrity/

Weiser, M. (1991). The computer for the 21 St Century. *Scientific American, 265*(3), 94–105. https://www.jstor.org/stable/24938718

Chapter 2
A Framework for Classifying Internet-of-Things Challenges

Richard B. Watson
https://orcid.org/0000-0002-5190-4662
Ryan Watson Consulting Pty. Ltd., Australia

Peter J. Ryan
https://orcid.org/0000-0002-0365-3192
Ryan Watson Consulting Pty. Ltd., Australia

ABSTRACT

The internet of things (IoT) is a global ecosystem of networked "things." It is the subject of much research worldwide, although it still has many challenges to overcome before it can achieve its full potential. Many papers have been written on the IoT and related areas including big data analytics, smart cities, and industrial IoT (IIoT). These challenges have mostly been seen as technical, although the IoT's business and societal challenges are also important. Most authors of research papers discuss the research challenges with which they are most familiar, but a framework which identifies and classifies all the challenges and cross-references the publications describing them in detail, is much needed. In this chapter, the authors extend their earlier IoT classification scheme to include more recent papers, and business and societal challenges as well as technical ones. The nature of the classification scheme and research challenges are described; however, the other chapters of this book cover in more detail the individual challenges and proposed strategies to mitigate them.

DOI: 10.4018/978-1-7998-9312-7.ch002

Figure 1. The Internet of Things showing that every 'thing' can be connected to the internet

INTRODUCTION

The Internet of Things (IoT) is a global ecosystem of "things" electronically connected by wired or wireless networks. One way of viewing the IoT is:

IoT = Human + Physical Objects (sensors, controllers, actuators, devices, computing storages) + Internet (Farhan et al., 2018) as shown in Figure 1.

However a consensus on a definition of IoT has not yet been reached, as enabling technologies keep evolving and new application domains are proposed (Ibarra-Esquer et al., 2017). Further, there are already emerging IoT subtypes such as Internet of Manufacturing Things, Internet of Medical Things, Internet of Military Things (IoMT), Ocean of Things, and Social Internet of Things that relate to specific application areas. Other authors include the Commercial Internet of Things (CIoT), the Consumer Internet of Things, and the Infrastructure Internet of Things (Anon, 2021). Many types of devices make up the IoT, including sensors, microcontrollers, transceivers, actuators and gateway-like devices (Nagasai, 2017). These come from numerous manufacturers worldwide, often using proprietary technologies, so that ensuring IoT devices can interoperate is one of the major challenges to be met.

Figure 2. Categories of IoT challenges

The IoT is often touted as a solution to the many problems besetting the world in the 21st century. These problems, including climate change, pandemics, environmental degradation, urbanization and military conflict, threaten the future of humanity on earth and are seen to be in urgent need of solution. Many IoT-based artifacts and systems have been produced to mitigate some of these problems, but much more remains to be done and many challenges remain in order for the IoT to reach its full potential. These challenges can be classified broadly as technical, business and societal, but these categories are linked together in complex ways as in Figure 2.

Plenty of research papers and books have been written on the IoT and related areas including big data analytics, smart cities and industrial IoT (IIoT). There are also many review papers which summarise and evaluate progress in IoT research and development reported in research papers, conferences and industry forums, so many in fact that several "reviews of the reviews" have been published (Aman et al., 2020; Swamy & Kota, 2020). As well, a growing number of review papers have been published on the topic of IoT challenges. Some of these restrict themselves to one particular set of challenges, such as privacy and security (Tawalbeh et al., 2020). Others set out to describe a number of challenges, although these papers have tended to confine themselves to technical challenges, or business challenges, or social challenges. Many authors concentrate on the particular challenges with which they are most familiar, which is understandable in such a complex field as the IoT. It must be noted however that a particular challenge, such as system-level design, is really a set of lower-level challenges, in this case including architecture, interoperability and scalability.

Interestingly, the papers cited here come from all parts of the world and not just the major research centres such as the US, Europe, and Northern Asia. Papers from the Middle East, South East Asia, Australia, and South America are also included. All countries including developing nations have investment in this new evolution of the internet that can stimulate their economic growth in areas including agriculture, transportation, utility management, and health (Alazie & Ebabye, 2019; Miazi et al., 2016).

The other chapters in this book describe in detail the individual IoT challenges and the progress being made to mitigate these. This chapter presents a framework which classifies all the challenges that the authors currently know of and cross-references key publications describing them in detail. It extends an earlier IoT classification scheme by the authors to include more recent papers, and business and societal challenges as well as technical ones (Ryan & Watson, 2017). Other papers on IoT challenges have proposed classification schemes which differ in some respects from ours, including the terminology used and the number of lower-level challenges considered. There is no one correct way to view these challenges, since the IoT ecosystem can be viewed as a technical system, a business system, or some other kind of system. It can perhaps be best described as a System-of-Systems (Schuck, 2021) and various systems thinking approaches can be used to tackle the complex challenges it presents (Ryan & Watson, 2017).

CLASSIFICATION SCHEME

Most recent surveys of the IoT include a section on research challenges, and the authors have attempted to consolidate their results for the purposes of this chapter. This is a difficult task due to differences in terminology by different authors, the fact that the different research challenges cannot be completely separated from each other, and the fact that they can be described at different levels of detail. For example, a very high-level research challenge might be "IoT design", but this includes a number of lower-level research challenges such as "architecture", "interoperability" and "scalability". Each of these lower-level research challenges may include other still lower-level research challenges.

Some authors consider IoT standardisation to be a research challenge in its own right, however the authors consider this to be a high-level research challenge which encompasses many lower level research challenges and so do not list it separately. In the authors' 2017 paper they divided these research challenges into the categories of Design, Scientific/Engineering and Management/Operations, although this is somewhat artificial, as several research challenges belong to more than one category

(Ryan & Watson, 2017). For example, Reliability/Robustness is a challenge at both the design and operational stages, as is Security/Privacy.

The scheme adopted is shown in Figure 3 from (Ryan & Watson, 2017). This scheme was focused on the technical challenges since operations research was its theme and did not include business and societal challenges that pose many issues.

Figure 3. Classification scheme used in (Ryan & Watson, 2017)

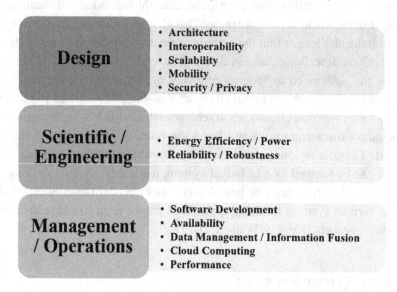

For detailed discussion of these technical research challenges, the reader is referred to the original references. Some challenges are only mentioned by a few papers, e.g., availability by Al-Fuqaha et al. (2015) and Cloud Computing by Gubbi et al. (2013) and Jain (2014). This is perhaps understandable as there are so many technologies which contribute to IoT that developing expertise in all of them is a considerable challenge. It is noted that a special issue of the journal *Computer Communications* in 2016 was entirely devoted to IoT research challenges, albeit predominantly technical (Borgia et al., 2016)).

In the present work the authors have expanded this classification scheme to include business and societal challenges. Further, they have given greater prominence to Big Data and its analysis and application. The updated classification scheme used in this chapter is shown as Figure 4.

Figure 4. Classification scheme adopted in present work

IOT TECHNICAL CHALLENGES

Many papers discuss technical challenges and these tend to include similar themes such as architecture, energy efficiency, security, interoperability, reliability, scalability, communications and so on. The IoT technical challenges broadly comprise the first three rows of the above figure – system level design challenges, management & operations challenges together with the new category of big data management challenges. This is displayed as Figure 5.

Figure 5. Technical challenges for IoT

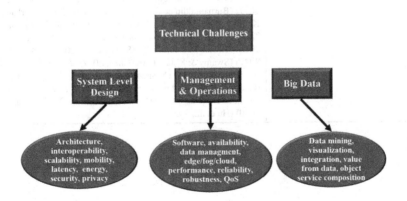

Table 1. System level design challenges

Challenge	Papers Referring
Architecture	(Borgia, 2014) (Stankovic, 2014) (Gubbi et al., 2013) (Chen et al., 2014) (Muralidharan et al., 2016) (Al-Fuqaha et al., 2015) (Motta et al., 2018) (Burhanuddin et al., 2017) (Singh et al., 2021) (Farhan & Kharel, 2019) (Han et al., 2016) (Nikoui et al., 2020) (Swamy & Kota, 2020) (Sen et al., 2018) (Georgakopoulos & Jayaraman, 2016) (Palattella et al., 2016) (Rayes & Salam, 2017) (Iqbal et al., 2020) (Lombardi et al., 2021)
Interoperability	(Borgia, 2014) (Jain, 2014) (Mattern & Floerkemeier, 2010) (Elkhodr et al., 2013) (Muralidharan et al., 2016) (Al-Fuqaha et al., 2015) (Triantafyllou et al., 2018) (Motta et al., 2018) (Giri et al., 2017) (Singh et al., 2021) (Farhan & Kharel, 2019) (Farhan et al., 2018) (Sisinni et al., 2018) (Nikoui et al., 2020) (Swamy & Kota, 2020) (Imran et al., 2020) (Korzun et al., 2013) (Sen et al., 2018) (Georgakopoulos & Jayaraman, 2016) (Naqvi et al., 2017) (Palattella et al., 2016) (Salman & Jain, 2017)
Scalability	(Jain, 2014) (Stankovic, 2014) (Mattern & Floerkemeier, 2010) (Al-Fuqaha et al., 2015) (Triantafyllou et al., 2018) (Motta et al., 2018) (Giri et al., 2017) (Burhanuddin et al., 2017) (Farhan & Kharel, 2019) (Farhan et al., 2018) (Nikoui et al., 2020) (Swamy & Kota, 2020) (Chen, 2012) (Imran et al., 2020) (Salman & Jain, 2017) (Chen & Zhang, 2014)
Mobility	(Borgia, 2014) (Mattern & Floerkemeier, 2010) (Al-Fuqaha et al., 2015) (Triantafyllou et al., 2018) (Kumar et al., 2016) (Swamy & Kota, 2020) (Palattella et al., 2016) (Salman & Jain, 2017) (Bouaziz & Rachedi, 2016)
Latency	(Sethi & Sarangi, 2017) (Swamy & Kota, 2020) (Sicari et al., 2020)
Energy Efficiency / Harvesting	(Jain, 2014) (Mattern & Floerkemeier, 2010) (Gubbi et al., 2013) (Chen et al., 2014) (Muralidharan et al., 2016) (Triantafyllou et al., 2018) (Burhanuddin et al., 2017) (Farhan & Kharel, 2019) (Farhan et al., 2018) (Kumar et al., 2016) (Sisinni et al., 2018) (Nikoui et al., 2020) (Swamy & Kota, 2020) (Chen, 2012) (Imran et al., 2020) (Korzun et al., 2013) (Sanislav et al., 2021)
Security/Privacy	(Borgia, 2014) (Jain, 2014) (Stankovic, 2014) (Mattern & Floerkemeier, 2010) (Elkhodr et al., 2013) (Gubbi et al., 2013) (Chen et al., 2014) (Muralidharan et al., 2016) (Al-Fuqaha et al., 2015) (Triantafyllou et al., 2018) (Motta et al., 2018) (Giri et al., 2017) (Burhanuddin et al., 2017) (Singh et al., 2021) (Farhan & Kharel, 2019) (Han et al., 2016) (Farhan et al., 2018) (*Internet of Things (IoT): Concepts and Applications*, 2020) (Kumar et al., 2016) (Tawalbeh et al., 2020) (Sisinni et al., 2018) (Nikoui et al., 2020) (Swamy & Kota, 2020) (Chen, 2012) (Imran et al., 2020) (Korzun et al., 2013) (Neshenko et al., 2019) (Sen et al., 2018) (Georgakopoulos & Jayaraman, 2016) (Naqvi et al., 2017) (Khan & Salah, 2018) (Pal, 2021) (HaddadPajouh et al., 2021) (Sicari et al., 2020) (Hu et al., 2021)

System Level Design Challenges

The system level design challenges and the key papers that refer to them are summarised in Table 1. These challenges are architecture, interoperability, scalability, mobility, latency, energy efficiency/harvesting, and security/privacy. Clearly there are many papers that discuss these issues.

Architecture

IoT architectures can use 3, 4, 5, 6 or more layers with 5 layers being the most common (Aman et al., 2020). The top layers refer to the application, the middle layers to the network functionality that handles data transmission, routing and processing across the IoT network, and the bottom layer to device connectivity. A more detailed generic architecture proposed by Triantafyllou et al. (2018) has 5 layers – the *perception/ device* layer, the *network layer* that transfers data from sensors to the processing system, the *middleware layer* that receives and processes data, the *application layer* that provides management of the objects resulting from the middleware layer, and the *business layer* that manages the whole IoT system. IoT-specific protocols such as Message Queuing Telemetry Transport (MQTT), Constrained Application Protocol (CoAP), or Advanced Messaging Queuing Protocol (AMQP) are employed at the application layer while communications standards such as WiFi, NB-IoT, Bluetooth Low Energy (BLE), Long Range Wireless Area Network (LoRaWAN) are employed at the network/middleware layers.

Addressing of devices in the IoT is required to uniquely identify each device. IPv4 uses 32-bit addresses and has already been depleted so that IPv6 that uses 128-bit addresses must be used (Borgia, 2014). IPv6 will enable all IoT devices to be accessible externally without the need for private addresses (Lombardi et al., 2021; Rayes & Salam, 2017; Walls, 2021).

Salman and Jain (2017) provide a comprehensive survey of protocols and standards that include communications, routing, network and session layers. These authors observe that the Institute of Electrical and Electronics Engineers (IEEE) standards focus on data link, the Internet Engineering Task Force (IETF) on networks and other organisations such as the International Telecommunications Union (ITU) work on security and management.

Interoperabilty

Interoperabilty refers to. the requirement for heterogenous devices to communicate with each other. These devices may use different communications standards, protocols, formats of data and even technologies since they are produced by different vendors and there is a lack of standards (Swamy & Kota, 2020). This was early noted as a key challenge for IoT systems by Borgia (2014) and Chen et al. (2014).

Scalability refers to the ability of the IoT to manage the increasing numbers of devices and the data these produce. With billions of devices already connected and many more being added constantly, the IoT needs to behave in the same manner for applications to operate efficiently. Lack of scalability will lead to poor Quality of Service (QoS) for applications such as Smart Cities.

Scalability is addressed by many authors including Chen (2012), Jain (2014), Stankovic (2014), Giri et al. (2017), Burhanuddin et al. (2017); Triantafyllou et al. (2018), Farhan et al. (2018), Motta et al. (2018), Farhan and Kharel (2019), Nikoui et al. (2020), Swamy and Kota (2020), Imran et al. (2020), and Salman and Jain (2017), and was much earlier identified as a challenge for Wireless Sensor Networks (Egea-López, 2006). The scalability challenge of storing big data is discussed by Chen and Zhang (2014).

Mobility

Mobility is the act of a node changing its attachment point due to changing topology that may result from issues such as poor network performance leading to delays and packet losses (Bouaziz & Rachedi, 2016). It can also refer to the physical movement of IoT devices in vehicles (cars, drones, etc.) and must also be considered since many IoT devices will be physically mobile.

Low levels of **latency** are essential for effective IoT operation. Swamy and Kota (2020) discuss latency reduction in terms of communication standards and protocols. They claim that MQTT and AMQP have lower latency than other protocols. The increasing deployment of 5G technology is driven in part by its lower latency (Sicari et al., 2020). Cloud computing may have too high latency for time critical applications requiring the inclusion of edge computing between the objects and cloud. They rate edge as low latency, fog as medium and cloud as high latency

Energy efficiency is vital for the success of the IoT. Communication and computation algorithms within IoT devices consume power so these algorithms need to be as efficient as possible. In addition many devices function in areas that are difficult to access and batteries cannot be easily replaced (Swamy & Kota, 2020). Further, there may be millions of such devices in a specific system such as a Smart City that will need to be powered continuously; replacing batteries in vast numbers of systems may not be practical. Most IoT devices are tiny and have limited space for larger batteries. Powering devices by harvesting environmental solar or RF radiation is one solution that is being investigated (Sanislav et al., 2021).

Table 2. Management/operations challenges

Challenge	Papers Referring
Software Development	(Mattern & Floerkemeier, 2010) (Muralidharan et al., 2016) (Motta et al., 2018) (Naqvi et al., 2017)
Availability	(Al-Fuqaha et al., 2015) (Sisinni et al., 2018) (Swamy & Kota, 2020)
Data Management/Fusion	(Borgia, 2014) (Jain, 2014) (Stankovic, 2014) (Mattern & Floerkemeier, 2010) (Elkhodr et al., 2013) (Gubbi et al., 2013) (Chen et al., 2014) (Triantafyllou et al., 2018) (Giri et al., 2017) (Farhan & Kharel, 2019) (Swamy & Kota, 2020) (Naqvi et al., 2017)
Edge/Fog/Cloud Computing	(Jain, 2014) (Gubbi et al., 2013) (Farhan & Kharel, 2019) (Swamy & Kota, 2020) (Chen, 2012) (Sen et al., 2018) (Hu et al., 2021) (Chen & Zhang, 2014)
Reliability/Robustness	(Borgia, 2014) (Stankovic, 2014) (Mattern & Floerkemeier, 2010) (Al-Fuqaha et al., 2015) (Farhan & Kharel, 2019) (Farhan et al., 2018) (Sisinni et al., 2018) (Nikoui et al., 2020) (Swamy & Kota, 2020) (Chen, 2012) (Imran et al., 2020) (Korzun et al., 2013) (Sarkar, 2016) (Qiu et al., 2016) (Xing, 2020)
Performance/QoS	(Borgia, 2014) (Gubbi et al., 2013) (Al-Fuqaha et al., 2015) (Motta et al., 2018) (Burhanuddin et al., 2017) (Sisinni et al., 2018) (Swamy & Kota, 2020)

Security and Privacy

IoT devices and systems need to be secure against cyber attack to protect data and ensure privacy. Security needs to be provided to all architecture layers; devices need to be authenticated before being added to the network; updates and patches need to be applied immediately. IoT devices will generally have limited CPU and memory so will be unable to employ sophisticated cyber defences. As the number of devices increases, so will the system vulnerability. Tawalbeh et al. (2020) provide an overview of privacy and security issues. For example, a typical IoT application will have a large number of identical devices (e.g. surveillance sensors) so that they will all have the same vulnerability level. The high number of interconnections also leads to security issues with each such connection adding a further channel of attack.

Recently much work has been done to mitigate IoT security and privacy challenges using Blockchain technology (Khan & Salah, 2018; Pal, 2021). The advantages and disadvantages of this technology in a 5G deployment are discussed by Sicari et al. (2020). Federated Learning is another recent approach to improve data security and privacy in a distributed IoT network (Hu et al., 2021). These approaches are described in detail in another chapter of this manual and other recent books published by IGI Global.

Management and Operations Challenges

Managing and operating IoT systems brings a new set of challenges due to the size and complexity of the IoT. The principal management and operational challenges for IoT and the key papers that refer to them are summarized in Table 2.

Software Development

Software development is rated as an IoT operational challenge by Mattern and Floerkemeier (2010), Muralidharan et al. (2016), Motta et al. (2018), and Naqvi et al. (2017). Mattern and Floerkemeier (2010) describe the extensive software development that will be required for managing smart devices that have minimal computation resources while still providing adequate support services There are also challenges for IoT operating systems to provide the networking and application development environments, for example tinyOS and contiki at the device level; and raspbian and ubuntu at the network level (Swamy & Kota, 2020).

Availability

Availability refers both to software availability and hardware availability that must be provided anywhere and at any time (Salman & Jain, 2017). Software availability provides services even when failures occur. Hardware availability provides easy access to devices that are compatible with the required protocols (Swamy & Kota, 2020). These software and hardware services need to be available to operate for long periods of time (Sisinni et al., 2018). Redundancy for critical devices and services may be required to ensure availability (Al-Fuqaha et al., 2015).

Data Management

Data management and fusion has been addressed by many authors. Borgia et al. (2016) claimed that the scale and heterogeneity of collected data leads to processing, analysis, and management issues. Management and storage of the vast quantity of IoT data is a huge issue (Triantafyllou et al., 2018).

Edge, fog, and Cloud Computing

Edge, fog, and cloud computing are three computational paradigms required to enable the IoT to operate efficiently and effectively.

Cloud Computing

Cloud computing refers to the process of using internet-hosted servers to store, manage and process data remotely rather than locally and has traditionally been used for IoT (Gubbi et al., 2013). Edge and fog computing are more recent developments.

Edge Computing

Edge computing is computation that happens at the network's edge, in close proximity to the physical location creating the data.

Fog Computing

Fog computing acts as a mediator between the edge and the cloud for various purposes, such as data filtering. Edge computing improves response and also provides low latency, security and privacy and high data aggregation (Sen et al., 2018; Swamy & Kota, 2020). A recent distributed machine learning approach termed Federated Learning enables greater data security and lower energy consumption by limiting data transmission between the edge and cloud (Hu et al., 2021). Fog computing works by utilising local devices termed fog nodes and edge devices. A straightforward example of edge and cloud computing is provided by Minella (2019) where an Arduino microcontroller manages the sensors, a Raspberry Pi microcomputer acts as the edge device and the data is streamed to Microsoft Azure.

Reliability

Reliability is the ability of each component in the IoT ecosystem to perform its essential tasks under different conditions. Each layer of the IoT architecture needs to have reliability for the whole system to perform correctly. Mission critical applications (such as healthcare related systems) will need more stringent reliability than less critical systems (such as home automation) (Sisinni et al., 2018). Xing (2020) discusses reliability models and solutions at four IoT layers – perception, communication, support, and application and claims that research in this area is immature with few solutions.

Robustness

Robustness is related to reliability. This is discussed by Swamy and Kota (2020) and also Sarkar (2016) and Qiu et al. (2016). These authors define robustness as the capacity of a network to provide and maintain an acceptable quality of service in the presence of faults. The IoT has many heterogenous networks and must be able to operate reliably with anticipated node failures.

Table 3. Big Data Management / Operations Challenges

Challenge	Papers Referring
Data Mining	(Marjani et al., 2017) (Chen, 2012) (Sunhare et al., 2020)
Data Visualization	(Marjani et al., 2017) (Protopsaltis et al., 2020) (Chen & Zhang, 2014)
Data Integration	(Bizer et al., 2012) (Alansari et al., 2018) (Papadokostaki et al., 2017) (Chen & Zhang, 2014)
Extracting Value from Data	(Bizer et al., 2012) (Chen & Zhang, 2014)
Object Service Composition	(Han et al., 2016) (Korzun et al., 2013) (Aoudia et al., 2019) (Hamzei & Navimipour, 2018)
Privacy	(Kumar et al., 2016) (Tawalbeh et al., 2020) (Marjani et al., 2017) (Korzun et al., 2013) (Papadokostaki et al., 2017) (Hu et al., 2021) (Pal, 2021)

Performance and QoS

Performance and QoS are related concepts. Performance can be measured from the rate of data collection, processing, and analysis. Performance of the IoT is critical for real time operations such as surveillance or health monitoring in hospitals. QoS refers to the requirement to deliver a satisfactory level of service to users. QoS measures can be used to describe the ability of the IoT to meet the user's needs. At the application level, Swamy and Kota (2020) define QoS ratings for different protocols.

Big Data Management / Operations Challenges

IoT systems can generate a huge volume, variety, and velocity of data. According to Desjardins (2019), 463 exabytes of data will be created each day by 2025 and much of this will be due to the IoT. This data needs to be collected, stored, analysed, visualized, and exploited by users. Data will be mostly unstructured and can be stored in NoSQL databases such as Cassandra. Systems such as the Apache freeware tools (Spark, Storm etc) and proprietary approaches can be used for data processing (Iqbal & Soomro, 2015). Machine learning is another approach to determine patterns in IoT big datasets. Sunhare et al. (2020) state that the data collected can be considered a new type of data known as IoT Big Data.

The Big Data management and operational challenges for IoT and the key papers that refer to these challenges are summarised in Table 3.

Data Mining

Data mining refers to the extraction of useful knowledge from masses of IoT data. Sunhare et al. (2020) reviewed the principal data mining techniques of classification where data objects are assigned to categories, clustering which classifies data objects into clusters with similar features, and association analysis that recognises patterns in and relationships among data, and other methods.

Data Visualization

Data visualization is a key part of IoT data analytics. Marjani et al. (2017) and Chen and Zhang (2014) describe how the size, heterogeneity and diversity of the data makes this a difficult task and most current visualization tools perform poorly. Protopsaltis et al. (2020) examined the tools, methodologies, and challenges for visualization of IoT data and reviewed systems such as PowerBI and Tableau. There are many python (matplotlib, seaborn, folium), R (ggplot2, plotly) and javaScript (D3.js, Angular) libraries for visualization. Augmented reality is another approach to visualize IoT data by providing the capability of monitoring the data in near real time. Virtual reality that immerses a user in a virtual world can also be applied for IoT visualization.

Data Integration

Data integration strategy is needed to allow for working with heterogeneous data sources. IoT devices transmit data in different formats and use different interfaces. The majority of IoT data will also be unstructured leading to ambiguities in interpretation. These data need to be integrated before meaningful results can be extracted. Challenges for Big Data integration are discussed by Bizer et al. (2012), Alansari et al. (2018), Papadokostaki et al. (2017) and Chen and Zhang (2014).

Much IoT data is unstructured or semi-structured and may include noise and redundancy.

Extracting Value

Extracting value from such data is not straightforward. (Papadokostaki et al., 2017) review the main approaches to handling IoT Big Data and the means of extracting value including the Apache Hadoop project and the semantic sensor web. Visualizing IoT data or using predictive Machine Learning are also ways to extract value from data. Chen and Zhang (2014) discuss extracting value from Big Data under the headings of data visualization and data curation.

Figure 6. Business and societal challenges

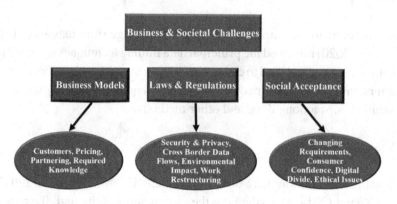

Service Composition

Han et al. (2016) describe service composition for smart objects. The IoT provides value from interaction of services from physical devices. Service composition of such smart objects is needed to support IoT applications. Aoudia et al. (2019) provided a review of approaches to service composition. They comment that insufficient work has been done in this critical area. Hamzei and Navimipour (2018) divide approaches to IoT service composition into the categories of framework, Service-Oriented Architecture (SOA) and RESTful, heuristic and model based.

Privacy and Security

Privacy and security are big challenges for IoT data. IoT systems generally have weak security protocols and policies according to Tawalbeh et al. (2020). The multilayer nature of IoT introduces security issues at each layer – data is sent from devices to the edge, fog and cloud with authorization and certification required at each computing platform. These authors propose an encryption approach to allow data to transfer between the various layers.

IOT BUSINESS AND SOCIETAL CHALLENGES

Business and societal challenges associated with the IoT are summarised in Figure 6. These comprise business models, laws & regulations, and social acceptance. These are discussed in the following sections.

Table 4. Business Model Challenges

Challenge	Papers Referring
Customers	(Hodapp et al., 2019) (Ju et al., 2016) (Fleisch, 2010) (Metallo et al., 2018) (Chan, 2015) (Turber et al., 2014) (Gassmann et al., 2014)
Pricing	(Sen et al., 2018) (Jindal et al., 2018) (Palattella et al., 2016) (Hodapp et al., 2019) (Ju et al., 2016) (Fleisch, 2010) (Metallo et al., 2018)
Partnering	(Hodapp et al., 2019) (Gloss, 2021b) (Ju et al., 2016) (Metallo et al., 2018; Turber et al., 2014) (Chan, 2016)
Required Knowledge	(Escribano et al., 2021) (Gloss, 2021b) (Kölsch et al., 2021) (Baig et al., 2019)

Business Models

The IoT is not simply an academic curiosity; it has the potential to transform the global economy through smart technology. End user spending on IoT solutions surpassed 200 $US billion in 2020 and may be nearly 2 trillion $US by 2025 according to Vailshery (2021). Klitou et al. (2017) described how Germany, for example, is promoting its 10-15 year *Industrie 4.0* strategy that aims to drive digitization throughout its economy.

The business model challenges and the key papers that refer to them are summarised in Table 4. These challenges are customers, pricing, partnering and required knowledge, and are discussed briefly below.

A business model is defined as the plan implemented by a company to generate revenue from operations and thus make a profit (Chan, 2016). Business models for the IoT ecosystem are not well understood and are still evolving. The basic elements are who, what, how and why: "who" is the customer, "what" is the value proposition, "how" is the value chain, and "why" is the underlying economic model (Gassmann et al., 2014).

Customers

Customers of IoT businesses may be other businesses or end consumers, and located in-country or anywhere across the globe. Business customers may be part of a vertical marketing system, in which producer, wholesaler and retailer collaborate to deliver products to end consumers. Alternatively, they may be part of a horizontal marketing system, in which businesses at the same level work together to gain economies of scale. Traditional business models are designed on a firm-centric basis; however due to the nature of the IoT ecosystem, in which firms must collaborate with competitors and across industries, it is easy to see why traditional business models are not adequate (Chan, 2016).

Pricing

Pricing of IoT products and services is difficult due to the need for firms to collaborate. However, price is an important element of a firm's business model. Pricing may be based on profit sharing, subscription fees or product sales (Ju et al., 2016). Furthermore, IoT platform providers may charge developers and/or enterprises, or offer their platforms free of charge (Hodapp et al., 2019).

Partnering

Partnering is important as the IoT ecosystem has many components: end-point providers, communications networks, base stations, cloud services, data analytics, etc. Individual firms must be capable of making smart collaborations, as collaborations form the fundamental basis of the IoT. If a firm is unable or unwilling to be collaborative, it will not be competitive (Chan, 2016). Many alliances exist at national and international levels, including the IoT Alliance Australia, LoRa Alliance, Mioty Alliance, and many others. Standards organisations including the International Organisation for Standardisation (ISO) and European Telecommunications Standards Institute (ETSI) facilitate such collaboration. Access to the partnering system may be open to all, proprietary, or a combination of these (Hodapp et al., 2019).

Required Knowledge

Another reason firms may choose to establish partnerships for IoT deployments is if they do not have the required knowledge in-house (Baig et al., 2019; Escribano et al., 2021). Many different skills are needed, including cybersecurity, edge computing, cloud, AI, data storage, applications and operational technology. Many of these skills are in very short supply. These partnerships may involve the use of consultants or outside experts, or outsourcing the entire IoT project to an IoT service provider that handles the whole deployment, from its design through its operation (Gloss, 2021b). Course providers, universities and vendors have developed more courses and certifications with a focus on IoT skills over the years. For example, the first Masters Degree in IoT in Australia has been established by La Trobe University, Bendigo Campus (Corner, 2019).

Table 5. Laws and regulations challenges

Challenge	Papers Referring
Security/Privacy	(Ryan & Watson, 2017) (Jindal et al., 2018) (Singh et al., 2018) (Brill & Jones, 2016) (Baldini et al., 2018) (Kobayashi et al., 2016) (Kumar et al., 2016) (Tawalbeh et al., 2020) (Sestino et al., 2020) (Palattella et al., 2016) (Gloss, 2021a) (AboBakr & Azer, 2017) (Singh et al., 2021) (Lee & Lee, 2015) (Kölsch et al., 2021) (Green, 2021) (Gilbert, 2017) (Khan & Salah, 2018) (Pal, 2021)
Cross Border Data Flows	(Jindal et al., 2018) (Singh et al., 2018) (Gloss, 2021a) (Baldini et al., 2018) (Pasquier et al., 2018) (Schuck, 2021)
Work Restructuring	(Ryan & Watson, 2017) (Sestino et al., 2020) (Gloss, 2021a) (Escribano et al., 2021) (ReliantVision, 2021) (Baig et al., 2019)
Environmental Impact	(Singh et al., 2021) (Sen et al., 2018) (Gloss, 2021a) (Solanki & Nayyar, 2019) (Vecchio, 2021) (Varjovi & Babaie, 2020) (Schuck, 2021)

Laws and Regulations

The challenges for laws and regulations and the key papers that refer to them are summarised in Table 5. These challenges are security/privacy, cross border data flows, work restructuring, and environmental input.

Laws define what is legally required in IoT devices, while regulations cover safety of end users and data privacy. Governments and standards bodies produce regulations, which vary at state, national & international levels. The United States and European regulations are described in detail in Gloss (2021a).

Security and Privacy

The security and privacy of IoT devices are the main foci of regulations. Device manufacturers must certify that their products meet regulations. Certification schemes standardise how device security is tested and define which functions to target.

Protection profiles provide security targets for device types. A basic protection profile requires IoT devices to use no universal password, secure interfaces and proven cryptography verified software with different progressive levels of security. There are optional protection levels, which give manufacturers a choice to meet the minimum security levels for a lower-cost product or reach higher security levels and advertise their product as being more secure. The IoT is a complex "system of systems", with components owned, managed, and operated by different people and organizations, perhaps in different geographies, with their own set of interests, incentives, responsibilities, and obligations (Singh et al., 2018). The IoT's scale exacerbates security issues given the vast numbers of components, their possible interconnections (all potential points of failure), and the many actors/vendors involved. Further, actuation, or failure to actuate, could result in physical harm. Active failure prevention and risk mitigation are important, as is auditing to facilitate learning from failures.

Collection of personal information may cause significant privacy risks to users. For example, the collected information may reveal the habits, locations, or physical conditions of an individual over time (Gilbert, 2017). The protection of personal data is a major focus of laws and regulations governing the IoT ecosystem. More than 120 countries have data-protection laws with broad application to businesses (Singh et al., 2018). European Union countries are covered by the General Data Protection Regulation (GDPR). The USA does not have a central federal level privacy law like the GDPR, although there are several vertically-focussed federal privacy laws, as well as state-based consumer-oriented privacy laws (Gloss, 2021a; Green, 2021). Further, the US Federal Trade Commission has a special tool for regulating IoT devices, the unfair tracking standard (Brill & Jones, 2016). In Australia the Privacy Act 1988, which includes the Australian Privacy Principles (APPs), is the principal data protection legislation. (Commonwealth of Australia, 2021).

Cross Border Data Flows

Cross border data flows are a consequence of the multi-national, multi-organizational nature of the IoT, and present challenges to regulators. There is often little visibility of, or means for control over, data once it is released to (accessed by) another party (Singh et al., 2018). Auditing flows of data can assist in demonstrating that the systems handling personal data satisfy regulatory and user requirements (Pasquier et al., 2018). It is also possible to manage data proactively as it moves across boundaries, using approaches such as sticky policies and information flow control (Singh et al., 2018).

Work Restructuring

The adoption of IoT and Big Data usually necessitates changes to business processes, business management and organizational culture. These have the potential to radically change the way businesses and people interact (Sestino et al., 2020). IoT projects can affect several departments in a company, from operational to business, and different departments must work together rather than as silos (Baig et al., 2019; Escribano et al., 2021). Further, the ability of the IoT to connect remote devices and people has facilitated businesses to expand their horizons and spread their reach across geographical boundaries. Implementing workplace restructuring is a challenging task,

Table 6. Social acceptance challenges

Challenge	Papers Referring
Changing Requirements	(Ryan & Watson, 2017) (Jindal et al., 2018) (Baldini et al., 2018) (Chan, 2016)
Consumer Confidence	(Ryan & Watson, 2017) (Jindal et al., 2018) (Kobayashi et al., 2016) (Palattella et al., 2016) (Baldini et al., 2018) (Liew et al., 2017) (Tsourela & Nerantzaki, 2020)
Digital Divide	(Baldini et al., 2018) (van Deursen et al., 2021)
Ethical Issues	(Sestino et al., 2020) (Vermanen et al., 2019) (Kobayashi et al., 2016) (Baldini et al., 2018) (AboBakr & Azer, 2017) (Lee & Lee, 2015)

but has the potential to enhance operational efficiency, boost employee productivity, increase workplace safety and reduce unnecessary expenses (ReliantVision, 2021).

Environmental Impact

The IoT presents many opportunities to minimise energy consumption, optimize the contributions of solar and wind power to the electrical grid, monitor air pollution and provide early warnings of threats such as pests, diseases, frosts, and droughts, before these become catastrophic events for crops (Vecchio, 2021) The Green Internet of Things (G-IoT) movement is making progress towards achieving these goals (Solanki & Nayyar, 2019; Varjovi & Babaie, 2020). On the other hand, the IoT can just as easily cause environmental problems due to the energy demands of billions of IoT devices, and the need to dispose of batteries and electronic waste. Environmental protection laws exist in most countries, although their degree of enforcement varies. Meteorological data from IoT sensors will be critical to monitoring national and international efforts to mitigate climate change in the years ahead.

Social Acceptance

The societal challenges of the IoT and the key papers that refer to them are summarised in Table 6. These challenges are changing requirements, consumer confidence, digital divide, and ethical issues and are discussed briefly below.

Changing Requirements

The possibilities created by digital technology in consumer electronics, transport, health and other application areas are causing customer requirements to change regularly. Smart connected devices are becoming more complicated as more and more devices do ever more complex jobs. This implies that companies must quickly adjust to market changes in order to succeed. Further, IoT devices are more like services than products. One's home may have its heating adjusted in anticipation of one arriving home, garage doors may open automatically, the home security system disarm, the doors unlock and the lights come on. Devices interact with each other, with the cloud and with people's smartphones. Other devices are constantly being added, removed or modified via downloads from the cloud. This implies customer service must focus less on product support and more on improved customer experience and lifetime value.

Consumer Confidence

New technology such as the IoT requires time for customers to accept and gain confidence. Some researchers have used the Technology Acceptance Model (TAM) to study this process (Liew et al., 2017). A model that reflects both psychological and technical aspects is needed to predict consumer behavioral intentions. The IoT is not just a new technology, but a completely new way for people to interact with their "brick-and-mortar" environments.

Digital Divide

Users have different sets of capabilities in accessing IoT devices and applications. Depending on their level of technical proficiency, users have different levels of perceptions of the privacy risks or different understandings of the requests sent to them through the IoT (Baldini et al., 2018). Those with higher education and higher incomes have more positive attitudes and are the first to actually buy IoT. This also means that they are first to develop the required skills and to engage in diverse IoT use. (van Deursen et al., 2021).

Ethical Issues

Ethical issues occur when a given decision, scenario or activity creates a conflict with a society's moral principles. The IoT and Big Data raise numerous ethical issues about privacy, responsibility and accountability for both individual employees and society (Vermanen et al., 2019). The correct identification of the author of the data collected by a typical IoT system is often hard to determine, as is the border between public and private life (AboBakr & Azer, 2017). The IoT can also damage trust in social relationships (Kobayashi et al., 2016).

FUTURE RESEARCH DIRECTIONS

The challenges described in this chapter are predominantly technical. There are many technical challenges still to be overcome for the device, architecture, and communications aspects although these are generally well understood. However future digital technology evolution is inevitable and will create new technical challenges. Big Data management and analysis is a critical area that warrants further investigation considering the explosive growth in data generated by the IoT.

Further research is needed in the business and societal aspects of the IoT as it becomes increasingly prevalent in all areas of society and business. Business models for the IoT are still under development and have not been adequately studied. Existing legal frameworks provide inadequate protection of privacy with billions of devices globally able to communicate via the internet and raise ethical issues

CONCLUSION

This chapter gives a brief overview of the IoT, challenges to its design, use and widespread adoption, and the important contemporary worldwide efforts to develop interoperability standards. These challenges have mostly been seen as technical, although business and societal challenges associated with the IoT are increasingly being recognised.

A framework for IoT challenges is proposed that includes technical, business, and societal aspects. This was developed by expanding the authors' previous framework for operations research challenges. The research for this chapter included review of a large number of journal articles, conference papers and industry reports, the numbers of which are rapidly increasing as the IoT has become a popular area both for academic research and business investment.

ACKNOWLEDGMENT

This research received no specific grant from any funding agency in the public, commercial, or not-for-profit sectors. The authors are principal consultants with Ryan Watson Consulting, a small Melbourne, Australia based company that works in the fields of Smart Cities, Big Data, and Internet of Things: https://www.ryanwatsonconsulting.com.au/.

REFERENCES

AboBakr, A., & Azer, M. A. (2017). *IoT ethics challenges and legal issues*. Paper presented at the 12th International Conference on Computer Engineering and Systems (ICCES 2017), Cairo, Egypt. Retrieved from https://www.researchgate.net/publication/322875867_IoT_ethics_challenges_and_legal_issues

Al-Fuqaha, A., Guizani, M., Mohammadi, M., Aledhari, M., & Ayyash, M. (2015). Internet of Things: A Survey on Enabling Technologies, Protocols, and Applications. *IEEE Communications Surveys and Tutorials*, *17*(4), 2347–2376. doi:10.1109/COMST.2015.2444095

Alam, M., Shakil, K. A., & Khan, S. (Eds.). (2020). *Internet of Things (IoT): Concepts and Applications*. Springer International Publishing. doi:10.1007/978-3-030-37468-6

Alansari, Z., Anuar, N. B., Kamsin, A., Soomro, S., Belgaum, M. R., Miraz, M. H., & Alshaer, J. (2018). Challenges of internet of things and big data integration. In *Lecture Notes of the Institute for Computer Sciences, Social Informatics and Telecommunications Engineering* (Vol. 200, pp. 47–55). Springer. doi:10.1007/978-3-319-95450-9_4

Alazie, G., & Ebabye, T. (2019). Impact of Internet of Thing in Developing Country: Systematic Review. *Internet of Things and Cloud Computing*, *7*(3), 65–72. doi:10.11648/j.iotcc.20190703.12

Aman, A. H. M., Yadegaridehkordi, E., Attarbashi, Z. S., Hassan, R., & Park, Y. (2020). A Survey on Trend and Classification of Internet of Things Reviews. *IEEE Access: Practical Innovations, Open Solutions*, *8*, 111763–111782. doi:10.1109/ACCESS.2020.3002932

Anon. (2021). *Internet of Things: The Five Types of IoT*. Retrieved from https://www.pareteum.com/internet-of-things-the-five-types-of-iot/

Aoudia, I., Benharzallah, S., Kahloul, L., & Kazar, O. (2019). Service composition approaches for Internet of Things: A review. *International Journal of Communication Networks and Distributed Systems*, *22*(1), 10017271. Advance online publication. doi:10.1504/IJCNDS.2019.10017271

Baig, M. I., Shuib, L., & Yadegaridehkordi, E. (2019). Big data adoption: State of the art and research challenges. *Information Processing & Management*, *56*(6), 102095. doi:10.1016/j.ipm.2019.102095

Baldini, G., Botterman, M., Neisse, R., & Tallacchini, M. (2018). Ethical design in the internet of things. *Science and Engineering Ethics, 24*(3), 905–925. doi:10.100711948-016-9754-5 PMID:26797878

Bizer, C., Boncz, P., Brodie, M. L., & Erling, O. (2012). The meaningful use of big data: Four perspectives — four challenges. *SIGMOD Record, 40*(4), 56–60. doi:10.1145/2094114.2094129

Borgia, E. (2014). The Internet of Things vision: Key features, applications and open issues. *Computer Communications, 54*, 1–31. doi:10.1016/j.comcom.2014.09.008

Borgia, E., Gomes, D. G., Lagesse, B., Lea, R., & Puccinelli, D. (2016). Special Issue on Internet of Things: Research Challenges and Solutions. *Computer Communications, 89-90*, 1–4. doi:10.1016/j.comcom.2016.04.024

Bouaziz, M., & Rachedi, A. (2016). A survey on mobility management protocols in Wireless Sensor Networks based on 6LoWPAN technology. *Computer Communications, 74*, 3–15. doi:10.1016/j.comcom.2014.10.004

Brill, H., & Jones, S. (2016). Little things and big challenges: Information privacy and the internet of things. *Am. UL Rev., 66*, 1183. doi:10.2139srn.3188958

Burhanuddin, M., Mohammed, A. A.-J., Ismail, R., & Basiron, H. (2017). Internet of things architecture: Current challenges and future direction of research. *International Journal of Applied Engineering Research: IJAER, 12*(21), 11055–11061.

Chan, H. C. Y. (2015). Internet of things business models. *Journal of Service Science and Management, 8*(4), 552–568. doi:10.4236/jssm.2015.84056

Chan, H. C. Y. (2016). Internet of things business models. In H. Geng (Ed.), *Internet of Things and Data Analytics Handbook* (pp. 735–757). Wiley. doi:10.1002/9781119173601.ch45

Chen, C. L. P., & Zhang, C.-Y. (2014). Data-intensive applications, challenges, techniques and technologies: A survey on Big Data. *Information Sciences, 275*, 314–347. doi:10.1016/j.ins.2014.01.015

Chen, S., Xu, H., Liu, D., Hu, B., & Wang, H. (2014). A vision of IoT: Applications, challenges, and opportunities with China perspective. *IEEE Internet of Things Journal, 1*(4), 349–359. doi:10.1109/JIOT.2014.2337336

Chen, Y.-K. (2012). *Challenges and opportunities of internet of things.*. doi:10.1109/ASPDAC.2012.6164978

Commonwealth of Australia. (2021). *Privacy Act 1988.* Federal Register of Legislation. Retrieved from https://www.legislation.gov.au/Details/C2021C00379

Corner, S. (2019). *La Trobe launches first Australian IoT masters course.* Retrieved from https://www.aumanufacturing.com.au/la-trobe-launches-first-australian-iot-masters-course

Desjardins, J. (2019). *How Much Data is Generated Each Day?* Retrieved from https://www.visualcapitalist.com/how-much-data-is-generated-each-day/

Egea-López, E., Vales-Alonso, J., Martinez-Sala, A., Pavon-Mario, P., & Garcia-Haro, J. (2006). Simulation scalability issues in wireless sensor networks. *IEEE Communications Magazine, 44*(7), 64–73. Advance online publication. doi:10.1109/MCOM.2006.1668384

Elkhodr, M., Shahrestani, S., & Cheung, H. (2013). *The Internet of Things: Vision & Challenges.* Paper presented at the TENCON Spring Conference.

Escribano, C. P., Theologou, N., Likar, M., Tryferidis, A., & Tzovaras, D. (2021). Business Models and Use Cases for the IoT. In C. Zivkovic, Y. Guan, & C. Grimm (Eds.), *IoT Platforms, Use Cases, Privacy, and Business Models* (pp. 51–80). Springer. doi:10.1007/978-3-030-45316-9_3

Farhan, L., & Kharel, R. (2019). Internet of Things: Vision, Future Directions and Opportunities. In S. C. Mukhopadhyay, K. P. Jayasundera, & O. A. Postolache (Eds.), *Modern Sensing Technologies* (pp. 331–347). Cham Springer International Publishing. doi:10.1007/978-3-319-99540-3_17

Farhan, L., Kharel, R., Kaiwartya, O., Quiroz-Castellanos, M., Alissa, A., & Abdulsalam, M. (2018). A concise review on Internet of Things (IoT)-problems, challenges and opportunities. In *2018 11th International Symposium on Communication Systems, Networks & Digital Signal Processing (CSNDSP)* (pp 1-6). IEEE.

Fleisch, E. (2010). What is the internet of things? An economic perspective. *Economics, Management, and Financial Markets,* (2), 125–157.

Gassmann, O., Frankenberger, K., & Csik, M. (2014). Revolutionizing the business model. In *Management of the fuzzy front end of innovation* (pp. 89–97). Springer International Publishing. doi:10.1007/978-3-319-01056-4_7

Georgakopoulos, D., & Jayaraman, P. P. (2016). Internet of things: From internet scale sensing to smart services. *Computing, 98*(10), 1041–1058. Advance online publication. doi:10.100700607-016-0510-0

Gilbert, F. (2017). Privacy and security legal issues. In H. Geng (Ed.), *Internet of Things and Data Analytics Handbook*. Wiley Telecom. doi:10.1002/9781119173601. ch43

Giri, A., Dutta, S., Neogy, S., Dahal, K., & Pervez, Z. (2017). Internet of things (IoT): a survey on architecture, enabling technologies, applications and challenges. In *Proceedings of the 1st International Conference on Internet of Things and Machine Learning*. ACM. 10.1145/3109761.3109768

Gloss, K. (2021a). *Navigate IoT regulations at local and global levels*. Retrieved from https://internetofthingsagenda.techtarget.com/feature/Navigate-IoT-regulations-at-local-and-global-levels?utm_campaign=20210928_Understand+IoT+legislation+at+the+state%2C+national+and+international+levels&utm_medium=EM&utm_source=NLN&track=NL-1843&ad=940157&asrc=EM_NLN_182707301

Gloss, K. (2021b). *Close the IoT skills gap with training and outside experts*. Retrieved from https://internetofthingsagenda.techtarget.com/feature/Close-the-IoT-skills-gap-with-training-and-outside-experts?utm_campaign=20210928_Understand+IoT+legislation+at+the+state%2C+national+and+international+levels&utm_medium=EM&utm_source=NLN&track=NL-1843&ad=940157&asrc=EM_NLN_182707307

Green, A. (2021). *Complete Guide to Privacy Laws in the US*. Retrieved from https://www.varonis.com/blog/us-privacy-laws/

Gubbi, J., Buyya, R., Marusic, S., & Palaniswami, M. (2013). Internet of Things (IoT): A vision, architectural elements, and future directions. *Future Generation Computer Systems*, *29*(7), 1645–1660. doi:10.1016/j.future.2013.01.010

HaddadPajouh, H., Dehghantanha, A., M. Parizi, R., Aledhari, M., & Karimipour, H. (2021). A survey on internet of things security: Requirements, challenges, and solutions. *Internet of Things*, *14*, 100129. doi:10.1016/j.iot.2019.100129

Hamzei, M., & Navimipour, N. J. (2018). Toward Efficient Service Composition Techniques in the Internet of Things. *IEEE Internet of Things Journal*, *5*(5), 3774–3787. doi:10.1109/JIOT.2018.2861742

Han, S. N., Khan, I., Lee, G. M., Crespi, N., & Glitho, R. H. (2016). Service composition for IP smart object using realtime Web protocols: Concept and research challenges. *Computer Standards & Interfaces*, *43*, 79–90. doi:10.1016/j.csi.2015.08.006

Hodapp, D., Remane, G., Hanelt, A., & Kolbe, L. M. (2019). *Business models for Internet of Things platforms: Empirical development of a taxonomy and archetypes.* Paper presented at the 14th International Conference on Wirtschaftsinformatik, Siegen, Germany.

Hu, K., Li, Y., Xia, M., Wu, J., Lu, M., Zhang, S., & Weng, L. (2021). Federated Learning: A Distributed Shared Machine Learning Method. *Complexity, 2021,* 1–20. Advance online publication. doi:10.1155/2021/8261663

Ibarra-Esquer, J. E., González-Navarro, F. F., Flores-Rios, B. L., Burtseva, L., & Astorga-Vargas, M. A. (2017). Tracking the evolution of the internet of things concept across different application domains. *Sensors (Basel), 17*(6), 1379. doi:10.339017061379 PMID:28613238

Imran, M. A., Zoha, A., Zhang, L., & Abbasi, Q. H. (2020). Grand Challenges in IoT and Sensor Networks. *Frontiers in Communications and Networks, 1*(7), 619452. Advance online publication. doi:10.3389/frcmn.2020.619452

Iqbal, M. A., Hussain, S., Xing, H., & Imran, M. A. (2020). *Enabling the Internet of Things: Fundamentals, Design and Applications.* John Wiley & Sons. doi:10.1002/9781119701460

Iqbal, M. H., & Soomro, T. R. (2015). Big data analysis: Apache storm perspective. *International Journal of Computer Trends and Technology, 19*(1), 9–14. doi:10.14445/22312803/IJCTT-V19P103

Jain, R. (2014). *Internet of Things: Challenges and Issues.* Paper presented at the 20th Annual Conference on Advanced Computing and Communications (ADCOM 2014), Bangalore, India. Retrieved from https://www.cse.wustl.edu/~jain/talks/iot_ad14.htm

Jindal, F., Jamar, R., & Churi, P. (2018). Future and challenges of internet of things. *International Journal of Computer Science and Information Technologies, 10*(2), 13–25. doi:10.5121/ijcsit.2018.10202

Ju, J., Kim, M.-S., & Ahn, J.-H. (2016). Prototyping Business Models for IoT Service. *Procedia Computer Science, 91,* 882–890. doi:10.1016/j.procs.2016.07.106

Khan, M. A., & Salah, K. (2018). IoT security: Review, blockchain solutions, and open challenges. *Future Generation Computer Systems, 82,* 395–411. doi:10.1016/j.future.2017.11.022

Klitou, D., Conrads, J., Rasmussen, M., Probst, L., & Pedersen, B. (2017). *Germany: Industrie 4.0*. https://ati.ec.europa.eu/sites/default/files/2020-06/DTM_Industrie%20 4.0_DE.pdf

Kobayashi, G., Quilici-Gonzalez, M. E., Broens, M. C., & Quilici-Gonzalez, J. A. (2016). The Ethical Impact of the Internet of Things in Social Relationships: Technological mediation and mutual trust. *IEEE Consumer Electronics Magazine*, *5*(3), 85–89. doi:10.1109/MCE.2016.2556919

Kölsch, J., Zivkovic, C., Guan, Y., & Grimm, C. (2021). An Introduction to the Internet of Things. In J. Kölsch, C. Zivkovic, Y. Guan, & C. Grimm (Eds.), *IoT Platforms, Use Cases, Privacy, and Business Models* (pp. 1–19). Springer. doi:10.1007/978-3-030-45316-9_1

Korzun, D., Balandin, S., & Gurtov, A. (2013). *Deployment of Smart Spaces in Internet of Things: Overview of the Design Challenges*. Paper presented at the 6th Conference on Internet of Things and Smart Spaces (ruSMART), St Petersburg, Russia. Retrieved from https://www.researchgate.net/publication/258994707_Deployment_ of_Smart_Spaces_in_Internet_of_Things_Overview_of_the_Design_Challenges

Kumar, P., Kunwar, R. S., & Sachan, A. (2016). A survey report on: Security & challenges in internet of things. *Proc National Conference on ICT & IoT*. Retrieved from https://www.academia.edu/20822175/A_Survey_Report_on_Security_and_ Challenges_in_Internet_of_Things

Lee, I., & Lee, K. (2015). The Internet of Things (IoT): Applications, investments, and challenges for enterprises. *Business Horizons*, *58*(4), 431–440. doi:10.1016/j. bushor.2015.03.008

Liew, C. S., Ang, J. M., Goh, Y. T., Koh, W. K., Tan, S. Y., & Teh, R. Y. (2017). Factors Influencing Consumer Acceptance of Internet of Things Technology. In I. R. M. Association (Ed.), *The Internet of Things: Breakthroughs in Research and Practice* (p. 16). IGI Global. doi:10.4018/978-1-5225-1832-7.ch004

Lombardi, M., Pascale, F., & Santaniello, D. (2021). Internet of Things: A General Overview between Architectures, Protocols and Applications. *Information (Basel)*, *12*(2), 87. doi:10.3390/info12020087

Marjani, M., Nasaruddin, F., Gani, A., Karim, A., Hashem, I., Siddiqa, A., & Yaqoob, I. (2017). Big IoT Data Analytics: Architecture, Opportunities, and Open Research Challenges. *IEEE Access: Practical Innovations, Open Solutions*, *5*, 5247–5261. Advance online publication. doi:10.1109/ACCESS.2017.2689040

Mattern, F., & Floerkemeier, C. (2010). From the Internet of Computers to the Internet of Things. In *From active data management to event-based systems and more* (pp. 242–259). Springer. doi:10.1007/978-3-642-17226-7_15

Metallo, C., Agrifoglio, R., Schiavone, F., & Mueller, J. (2018). Understanding business model in the Internet of Things industry. *Technological Forecasting and Social Change, 136*, 298–306. Advance online publication. doi:10.1016/j.techfore.2018.01.020

Miazi, N. S., Erasmus, Z., Razzaque, M. A., Zennaro, M., & Bagula, A. (2016). Enabling the Internet of Things in developing countries: Opportunities and challenges. In *2016 International Conference on Informatics, Electronics and Vision (ICIEV)* (pp 564-569). 10.1109/ICIEV.2016.7760066

Minella, M. (2019). *Arduino, Raspberry and Azure IoT Edge - a practical implementation.* Retrieved from https://www.linkedin.com/pulse/arduino-raspberry-azure-iot-edge-practical-mauro-minella

Motta, R., Oliveira, K., & Travassos, G. (2018). *On challenges in engineering IoT software systems.* Paper presented at the XXXII Brazilian Symposium on Software Engineering, SBES 2018, Sao Carlos, Brazil. Retrieved from https://www.researchgate.net/publication/327651553_On_challenges_in_engineering_IoT_software_systems

Muralidharan, S., Roy, A., & Saxena, N. (2016). An Exhaustive Review on Internet of Things from Korea's Perspective. *Wireless Personal Communications, 90*(3), 1–24. doi:10.100711277-016-3404-8

Nagasai. (2017). *Classification of IoT Devices.* Retrieved from https://www.cisoplatform.com/profiles/blogs/classification-of-iot-devices

Naqvi, S., Hassan, S., & Hussain, F. (2017). IoT Applications and Business Models. In *Internet of Things* (pp. 45–61). Building Blocks and Business Models. doi:10.1007/978-3-319-55405-1_4

Neshenko, N., Bou-Harb, E., Crichigno, J., Kaddoum, G., & Ghani, N. (2019). Demystifying IoT security: An exhaustive survey on IoT vulnerabilities and a first empirical look on Internet-scale IoT exploitations. *IEEE Communications Surveys and Tutorials, 21*(3), 2702–2733. doi:10.1109/COMST.2019.2910750

Nikoui, T. S., Rahmani, A. M., & Balador, A. (2020). Internet of Things architecture challenges: A systematic review. *International Journal of Communication Systems, 34*(4), e4678. doi:10.1002/dac.4678

Pal, K. (2021). Blockchain Technology With the Internet of Things in Manufacturing Data Processing Architecture. In A. B. Mnaouer & L. C. Gourati (Eds.), *Enabling Blockchain Technology for Secure Networking and Communications* (pp. 229–247). IGI Global. doi:10.4018/978-1-7998-5839-3.ch010

Palattella, M. R., Dohler, M., Grieco, A., Rizzo, G., Torsner, J., Engel, T., & Ladid, L. (2016). Internet of Things in the 5G Era: Enablers, Architecture, and Business Models. *IEEE Journal on Selected Areas in Communications*, *34*(3), 510–527. doi:10.1109/JSAC.2016.2525418

Papadokostaki, K., Mastorakis, G., Panagiotakis, S., Mavromoustakis, C. X., Dobre, C., & Batalla, J. M. (2017). Handling big data in the era of internet of things (IoT). In *Advances in mobile Cloud computing and Big Data in the 5G Era* (pp. 3–22). Springer. doi:10.1007/978-3-319-45145-9_1

Pasquier, T., Singh, J., Powles, J., Eyers, D., Seltzer, M., & Bacon, J. (2018). Data provenance to audit compliance with privacy policy in the Internet of Things. *Personal and Ubiquitous Computing*, *22*(2), 333–344. doi:10.100700779-017-1067-4

Protopsaltis, A., Sarigiannidis, P., Margounakis, D., & Lytos, A. (2020). Data Visualization in Internet of Things: Tools, Methodologies, and Challenges. *ARES 2020: Proceedings of the 15th International Conference on Availability, Reliability and Security*. doi: 10.1145/3407023.3409228

Qiu, T., Luo, D., Xia, F., Deonauth, N., Si, W., & Tolba, A. (2016). A greedy model with small world for improving the robustness of heterogeneous Internet of Things. *Computer Networks*, *101*, 127–143. doi:10.1016/j.comnet.2015.12.019

Rayes, A., & Salam, S. (2017). IoT Protocol Stack: A Layered View. In Internet of Things From Hype to Reality: The Road to Digitization (pp 93-138). Springer International Publishing. doi:10.1007/978-3-319-44860-2_5

ReliantVision. (2021). *The Role of IoT in Restructuring the Workplace*. Retrieved from https://reliantvision.com/the-role-of-iot-in-workplace/

Ryan, P. J., & Watson, R. B. (2017). Research Challenges for the Internet of Things: What Role Can OR Play? *Systems*, *5*(1), 24. doi:10.3390ystems5010024

Salman, T., & Jain, R. (2017). A Survey of Protocols and Standards for Internet of Things. *Advanced Computing and Communications*, *1*. Advance online publication. doi:10.34048/2017.1.F3

Sanislav, T., Mois, G. D., Zeadally, S., & Folea, S. C. (2021). Energy Harvesting Techniques for Internet of Things (IoT). *IEEE Access: Practical Innovations, Open Solutions, 9*, 39530–39549. doi:10.1109/ACCESS.2021.3064066

Sarkar, S. (2016). Internet of Things—Robustness and reliability. In R. Buyya & A. V. Dastjerdi (Eds.), *Internet of Things* (1st ed., pp. 201–218). Elsevier. doi:10.1016/B978-0-12-805395-9.00011-3

Schuck, T. M. (2021). *Cybernetics, Complexity, and the Challenges to the Realization of the Internet-of-Things*. Paper presented at the Complex Adaptive Systems Conference, Malvern, PA. Retrieved from https://www.sciencedirect.com/science/article/pii/S187705092101125X

Sen, J., Lee, M., Lee, S., Choe, Y., Domb, M., Pal, A., ... Mihovska, A. (2018). *Internet of Things: Technology, Applications and Standardardization*. INTECH Publishers; doi:10.5772/intechopen.70907

Sestino, A., Prete, M. I., Piper, L., & Guido, G. (2020). Internet of Things and Big Data as enablers for business digitalization strategies. *Technovation, 102173*. Advance online publication. doi:10.1016/j.technovation.2020.102173

Sethi, P., & Sarangi, S. (2017). Internet of Things: Architectures, Protocols, and Applications. *Journal of Electrical and Computer Engineering, 2017*, 1–25. doi:10.1155/2017/9324035

Sicari, S., Rizzardi, A., & Coen-Porisini, A. (2020). 5G In the internet of things era: An overview on security and privacy challenges. *Computer Networks, 179*, 107345. doi:10.1016/j.comnet.2020.107345

Singh, D., Jerath, H. P. R., Tripathi, S. L., & Sanjeevikumar, P. (2021). Internet of Things – Definition, Architecture, Applications, Requirements and Key Research Challenges. In D. Singh, H. Jerath, R. P, S. Tripathi, & P. Sanjeevikumar (Eds.), Design and Development of Efficient Energy Systems (pp 285-295). Scrivener Publishing LLC. doi:10.1002/9781119761785.ch16

Singh, J., Millard, C., Reed, C., Cobbe, J., & Crowcroft, J. (2018). Accountability in the IoT: Systems, law, and ways forward. *Computer, 51*(7), 54–65. doi:10.1109/MC.2018.3011052

Sisinni, E., Saifullah, A., Han, S., Jennehag, U., & Gidlund, M. (2018). Industrial Internet of Things: Challenges, Opportunities, and Directions. *IEEE Transactions on Industrial Informatics*. . doi:10.1109/TII.2018.2852491

Solanki, A., & Nayyar, A. (2019). Green internet of things (G-IoT): ICT technologies, principles, applications, projects, and challenges. In G. Kaur & P. Tomar (Eds.), Handbook of Research on Big Data and the IoT (pp. 379-405). IGI Global.

Stankovic, J. A. (2014). Research directions for the internet of things. *Internet of Things Journal, IEEE, 1*(1), 3–9. doi:10.1109/JIOT.2014.2312291

Sunhare, P., Chowdhary, R. R., & Chattopadhyay, M. K. (2020). Internet of things and data mining: An application oriented survey. *Journal of King Saud University – Computer and Information Sciences.* . doi:10.1016/j.jksuci.2020.07.002

Swamy, S. N., & Kota, S. R. (2020). An Empirical Study on System Level Aspects of Internet of Things (IoT). *IEEE Access: Practical Innovations, Open Solutions, 8*, 188082–188134. doi:10.1109/ACCESS.2020.3029847

Tawalbeh, L., Muheidat, F., Tawalbeh, M., & Quwaider, M. (2020). IoT Privacy and Security: Challenges and Solutions. *Applied Sciences (Basel, Switzerland), 10*(12), 4102. doi:10.3390/app10124102

Triantafyllou, A., Sarigiannidis, P., & Lagkas, T. D. (2018). Network Protocols, Schemes, and Mechanisms for Internet of Things (IoT): Features, Open Challenges, and Trends. *Wireless Communications and Mobile Computing, 24*, 1–24. Advance online publication. doi:10.1155/2018/5349894

Tsourela, M., & Nerantzaki, D.-M. (2020). An Internet of Things (IoT) Acceptance Model. Assessing Consumer's Behavior toward IoT Products and Applications. *Future Internet, 12*(11), 191. doi:10.3390/fi12110191

Turber, S., Brocke, J. v., Gassmann, O., & Fleisch, E. (2014). *Designing Business Models in the Era of the Internet of Things.* Paper presented at the 9th International Conference on Design Science Research in Information Systems and Technology (DESRIST 2014), Miami, FL. Retrieved from https://www.researchgate.net/publication/265166080_Designing_Business_Models_in_the_Era_of_the_Internet_of_Things

Vailshery, L. S. (2021). *Forecast end-user spending on IoT solutions worldwide from 2017 to 2025 (Statista).* Retrieved from https://www.statista.com/statistics/976313/global-iot-market-size/#:~:text=The%20global%20market%20for%20Internet%20of%20things%20%28IoT%29,1.6%20trillion%20by%202025.%20The%20Internet%20of%20Things van Deursen, A. J., van der Zeeuw, A., de Boer, P., Jansen, G., & van Rompay, T. (2021). Digital inequalities in the Internet of Things: Differences in attitudes, material access, skills, and usage. *Information Communication and Society, 24*(2), 258–276. doi:10.1080/1369118X.2019.1646777

Varjovi, A. E., & Babaie, S. (2020). Green Internet of Things (GIoT): Vision, applications and research challenges. *Sustainable Computing: Informatics and Systems, 28*, 100448. doi:10.1016/j.suscom.2020.100448

Vecchio, M. (2021). Guest Editorial: IoT and the Environment. *IEEE Internet of Things Magazine, 4*(1), 10–11. doi:10.1109/MIOT.2021.9390452

Vermanen, M., Rantanen, M. M., & Harkke, V. (2019). *Ethical challenges of IoT utilization in SMEs from an individual employee's perspective.* Paper presented at the 27th European Conference on Information Systems (ECIS), Stockholm, Sweden. Retrieved from https://aisel.aisnet.org/ecis2019_rp/112

Walls, J. (2021). *Why move to an all-IP IoT?* Retrieved from https://internetofthingsagenda.techtarget.com/post/Why-move-to-an-all-IP-IoT

Xing, L. (2020). Reliability in Internet of Things: Current Status and Future Perspectives. *IEEE Internet of Things Journal.* . doi:10.1109/JIOT.2020.2993216

APPENDIX: ACRONYMS

Table 7.

AI	Artificial Intelligence
AMQP	Advanced Messaging Queuing Protocol
BLE	Bluetooth Low Energy
CIoT	Commercial Internet of Things
CoAP	Constrained Application Protocol
ETSI	European Telecommunications Standards Institute
GDPR	General Data Protection Regulation
G-IoT	Green Internet of Things
IEEE	Institute of Electrical and Electronics Engineers
IETF	Internet Engineering Task Force
IIoT	Industrial Internet of Things
IoMT	Internet of Military Things
IoT	Internet of Things
IPv4	Fourth version of the Internet Protocol (IP)
IPv6	Sixth version of the Internet Protocol
ISO	International Organisation for Standardisation
ITU	International Telecommunications Union
LoRa	Long Range - an IoT Physical Layer Implementation
LoRaWAN	Long Range Wireless Area Network
Mioty	Fraunhofer-developed IoT Physical Layer Implementation
MQTT	Message Queuing Telemetry Transport
QoS	Quality of Service
REST	REpresentational State Transfer Architecture
TAM	Technology Acceptance Model

Chapter 3
Challenges of Managing IoT Networks and Prospective Measures

Anjum Sheikh Qureshi
Rajiv Gandhi College of Engineering Research and Technology, Chandrapur, India

ABSTRACT

Internet of things (IoT) is an evolving technology that has interconnected devices and humans across the world. The number of IoT users and devices is growing rapidly. The versatile nature of IoT has led to the development of many new applications. Establishing communication among the interconnected devices at any time from any place is the primary objective of IoT. With the increasing usage of IoT devices, it has become difficult to handle them and monitor their communication process. Some of the challenges faced by the IoT networks are data management, energy consumption, connectivity, security, and addressing of devices. It is essential to overcome these challenges to increase the popularity and acceptance level among the existing users and convince others to adopt IoT into their daily lives.

INTRODUCTION

Internet of Things (IoT) is a network of many interconnected objects, devices, humans, and services that can communicate with each other and exchange data. usage of IoT devices has increased rapidly in the last decade and it has been predicted that the number of IoT devices will surpass the global population in the coming years. IoT is being used in different domains like health care, agriculture, transportation,

DOI: 10.4018/978-1-7998-9312-7.ch003

and many more. Despite the fast growth of IoT users and its adoption, managing IoT networks remains a big challenge. It is essential to overcome the challenges as many IoT initiatives will fail to meet the expectations of their users.

Some of the challenges while managing the IoT networks are management of data, energy consumption, privacy and security of data, connectivity, and management of the devices. IoT involves a large amount of data that is being exchanged among the devices through the networks. The data has to be stored, processed, analyzed, and transferred to their destinations to facilitate the working of IoT devices and their related applications. The IoT devices due to limited storage capacities are unable to store and process the voluminous data.

Another challenge is the battery-powered IoT devices or nodes with limited power capabilities. The power level of the IoT nodes is required to maintain communication among the devices. Failure of any one of the nodes can disrupt the whole route and would not allow the communication to be completed. The storage technologies are improving but as the battery power is limited energy consumption remains one of the main challenges for IoT devices. The IoT devices or the nodes are generally small in size due to which the batteries are small and therefore it is not possible to have batteries with a large power capacity. The third challenge is the connectivity of IoT devices. Some of the IoT devices can work on wired Ethernet connections while most of them depend on wireless communication technologies. IoT communications mainly rely on the internet for the communication or the exchange of information over the network. Internet connectivity is thus a requirement for most IoT communications. In remote areas or places located at adverse geographical locations, it is difficult to carry communications with the internet.

The large amount of data involved with IoT makes it susceptible to breaching and hacking and therefore security and privacy of data is a great concern for the communications of the IoT platform. It is important to avoid the connection of unauthorized devices from connecting to the network. An authentication framework is a must for IoT devices to reduce the risks of hacking. Most of the IoT applications are used by people in day-to-day life. If these applications get hacked it can be fatal for the people. Management of IoT devices is another challenge that needs attention to increase the adaptability of the technology among its users. Several components are involved in completing the IoT network or enabling communication among the devices. Sensors, actuators, controllers, etc have to be used but all these devices have to be installed, configured, and monitored for ensuring proper performance. In addition to this, these devices have to be updated regularly so that they can address the problems that may arrive during communication. The interoperability and heterogeneous nature of the IoT make it more difficult to manage the devices. Looking at the problems it is, therefore, necessary to devise solutions so that the service providers, organizations, and the users are satisfied and use IoT to make

Figure 1. Challenges of IoT
(Farhan et.al 2018)

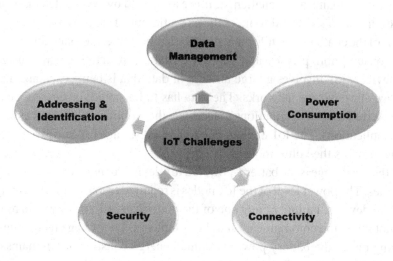

their life easier. This chapter will discuss in detail the challenges mentioned above and suggest solutions to deal with them.

The chapter will be organized as follows: section 1 will introduce the chapter, section 2 will discuss the first challenge that is data management and its possible solutions, section 3 will highlight the power consumption issues and methods that can be adopted to increase the lifetime of IoT networks by reducing it. Section 4 will shed light on the connectivity technologies used in IoT, the importance of connectivity in IoT, challenges, and some measures that can be adopted to solve the issues of connectivity. Section 5 will discuss the data privacy and security issues and measures to minimize the risks of breaching in the IoT framework. Section 6 will talk about identification issues faced by IoT devices and explain methods to deal with these issues. Section 7 will conclude the chapter.

DATA MANAGEMENT OF IoT

Internet of Things (IoT) consists of many smart objects that continuously generate data and transfer it over the internet. The current research in the field of IoT has contributed to lowering the manufacturing costs, improving the energy efficiency of IoT devices and communication technologies to enable interconnectivity of the devices. At the same time, there are solutions to manage the voluminous data of IoT (Abu-Elkheir et al.,2013). The sensors of the IoT devices used by the consumers gather data from wearables, smart appliances, security systems, traffic monitoring systems, etc. Few types of data collected by the IoT devices include automation data that is collected from the automated devices, status data that is collected as raw data and then used for analysis, and location data that helps in tracking the devices in real-time. This data has to be stored and processed before using it. The massive data thus associated with the numerous devices pose a challenge for its storage due to the limited storage capacities of the IoT devices. The users will be able to derive more benefits from the IoT if there are platforms that can store as well as process the data. Due to the voluminous data that is predicted to increase in the coming years and the processing and analysis required to use the data as the information, it is required that there should be unlimited storage space. In addition to this, the storage platform should be able to process the data efficiently to avoid error-free transmission. The solutions available for the storage and handling of IoT data are cloud computing and fog computing.

Cloud Computing

Cloud computing is a paradigm that enables on-demand network access to the shared pool of computing resources like networks, storage, services, applications, and servers. It is a highly scalable and reliable platform that allows sharing of resources by many users. Many organizations prefer the cloud as it provides unlimited storage for their data, keeps it secure, and provides universal access. The investments required for the data processing and management are minimized as the organizations do not require big data centers for the data (Mell & Grance, 2009). The different types of cloud deployment models as defined by the National Institute of Standards and Technology (NIST) include public, private, hybrid, and community (Rountree & Castrillo, 2014). A public cloud is owned by an external service provider that can be used by anyone by paying for the services. A private cloud on the other hand is owned by a single organization that can control the purchase, maintenance, and support of the cloud services. The hybrid cloud combines the properties of both public and private cloud by which the organizations can derive the benefits of both deployment models. Organizations can move the data and applications among the private and public cloud depending on the actions that need to be performed. The community cloud is set up by few organizations that work for a similar purpose. They share the resources, maintenance, support, and investment needed for the infrastructure. The three types of service models of cloud are Software as a Service (SaaS), Platform as a Service (PaaS), and Infrastructure as a Service (IaaS) (Laszewski & Nauduri, 2012). Some examples of SaaS are emails, google apps, and social media platforms. In SaaS, the end-users can use the applications by using either a web browser or a program interface. In PaaS, the organizations can develop and manage applications that can be offered as services by using the internet. The resources needed for the

Figure 2. Service models of Cloud (a) SaaS (b) Paas (c) IaaS
(Puthal et.al, 2015)

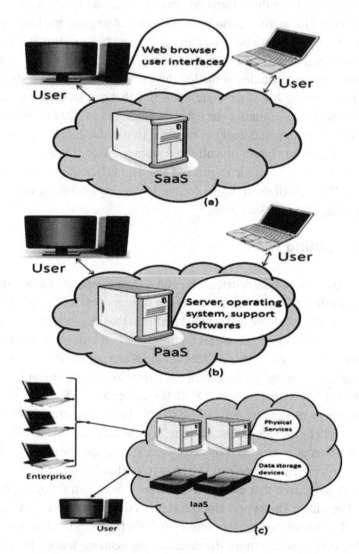

processing data and the infrastructure components like storage, network devices, and firewall are supplied by the IaaS.

Some of the features of cloud computing due to which the companies have migrated towards cloud are (Mell & Grance,2009; Sasikala, 2013; Olive 2012):

- **On-Demand Self Service**: The consumer can access the resources and the computing capabilities without human intervention. Some of the public providers like Amazon, Google, and Microsoft allow modification of services by the client organization without interacting with the hosting provider. The level of modifications is determined by the service level agreement (SLA) to prevent the request for unduly capacity changes.

- **Broad Network Access:** The hosted applications are available over the network that can be accessed through gadgets like smartphones, tablets, computers, and laptops. These devices use built-in browsers to attain broad network access.
- **Resource Pooling:** The computing resources of the service providers are pooled so that they can serve multiple consumers at a time according to their demands. The companies can purchase more resources without investing in the physical infrastructure. Organizations of similar security levels are grouped in specific cloud infrastructures. For example, all the pharmaceutical organizations and all the federal organizations will be placed on separate cloud infrastructure.
- **Rapid Elasticity:** It can handle the changes in demand of the consumers either semi-automatically or sometimes automatically. The capabilities of the cloud can be extended to fulfill the anticipated hikes in the usage of the cloud due to the increase in the number of users. Due to this, a cloud appears like a storage space with unlimited capacities to the users that can be purchased by them at any time according to their requirements.
- **Measured Service:** The analysis of the data required by the users is performed automatically. It supports a metering service by which the users can monitor, control and report the usage of the cloud. This system provides transparency of the utilized services to the service providers as well as the consumers.

Fog Computing

Fog computing can be defined as an extension of the cloud computing paradigm that consists of man edge nodes directly connected to the devices. It is, therefore, closer to the users as compared to the centralized data centers of the cloud. Fog computing enables a new category of application and services by extension of cloud computing to the edge of the network. It is a source of fruitful interaction with the cloud to facilitate the management and analytics of the data generated by the IoT devices (Bonomi, et. al, 2012). The cloud computing platform can store and process the enormous IoT data at reduced costs but suffers from the problem of high latency and security and privacy of data. Most IoT applications require low latency but due to the distance between the client and the data centers, the latency tends to increase. Secondly, as the private data is transferred through globally connected channels the cloud data is vulnerable to security attacks and loss of data. This problem can be solved partially by using hybrid and private clouds.

Figure 3. Fog computing architecture
(Cha et.al 2018)

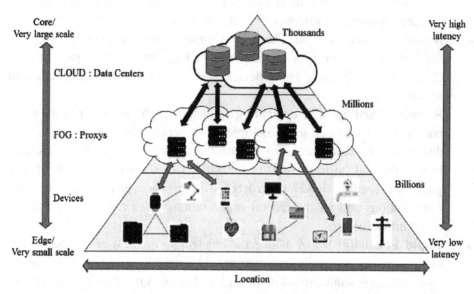

The fog computing layer acts as an intermediate between the devices and the cloud server. It solves the problem of high latency faced by the cloud by regulating the data that has to be sent to the cloud and that can be managed locally (Iorga et.al, 2018). Some of the benefits of the closeness of the fog with the devices as described by (Varghese et.al, 2018) are that it improves the mobility of data due to the direct communication with the devices and enables real-time interaction rather than the batch processing used in cloud computing. Fog has its security implementation by which it handles the protection and privacy issues faced by the cloud (Wadhwa & Aron, 2018). Its decentralized architecture and location awareness enables delivering of fine quality streaming services.

ENERGY CONSUMPTION ISSUES

The energy constraints of the IoT devices increase the challenges of the interconnection of things in an interoperable manner due to the energy requirements of the communications on the network. Low power communication technologies and energy harvesting techniques are some of the solutions to deal with the energy issues of low-power IoT devices.

Less Energy-Consuming Communication Technologies

Some of the commonly used low power communication technologies that can be used for IoT are:

- **IEEE 802.15.4:** It is a protocol that consumes less energy. It is considered to be a good option for a low data rate scenario and provides a nominal data rate of 250 kbps that is based on offset QPSK (Quadrature Phase Shift Keying) modulation. The protocol is simple, can be used for low to medium range communication. Due to these reasons, it is suitable for power-constrained IoT devices (Subrahmanyam et.al, 2018).
- **Bluetooth Low Energy (BLE):** It is another wireless standard based on GFSK (Gaussian Frequency Shift Keying) modulation that is considered suitable for ultra low power IoT applications. It offers a nominal data rate of 1Mbps which is higher than IEEE 802.15.4. Due to the higher data rate, it requires less time to transfer the data frames and therefore consumes lesser energy than the IEEE 802.15.4 (Fafoutis et al., 2016).
- **Ultra-Wideband (UWB) Technology:** UWB is another technology that provides high-speed communication but consumes less power. It uses carrier less transmission of data in the form of short pulses which makes it suitable for short-range applications (Sharma et.al, 2020). The UWB based devices are capable of handling the demanding needs of connectivity of the power-constrained IoT devices due to their wide bandwidth and high-speed communication (Kirtania et.al, 2021).
- **Dash 7:** It is a simple, low-power, wireless communication protocol suitable for IoT requirements. Due to its high data rate, it is used for designing scalable, long-range outdoor coverage applications. It is a low cost and versatile technology with less latency due to which it is considered a favorable solution for applications that require low power consumption (Cetinkaya 2015)
- **RFID:** Radio Frequency Identification (RFID) is one of the key technologies that are used for IoT applications because of its properties like high efficiency and low power consumption. It is a wireless communication technology that supports non-contact reading and writing at a distance of few centimeters to few meters. It can recognize high speed moving objects, can identify multiple targets concurrently, and provide strong security (Chen & Jin, 2012)

Energy Harvesting

Using cable-powered devices for IoT is not considered to be feasible for IoT devices due to its high cost and difficulty in its deployment. The IoT devices are powered by

Figure 4. An energy harvesting system
(Elahi et.al 2020)

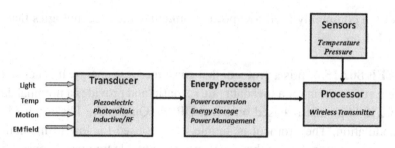

batteries that have a limited lifetime. Battery replacements can be effective for small IoT systems but are expensive and impracticable for large systems. As a result, there is a need to adopt alternative energy sources for independent and large-scale IoT applications. Energy harvesting techniques are used nowadays to increase the lifetime of IoT devices. It is a technique in which the energy from any renewable source is collected and converted into electrical energy. An energy harvesting system that uses light, motion, temperature, and electromagnetic field has been given in fig 4.

Some of the sources of energy harvesting are solar energy, wind energy, thermal energy, mechanical energy, and radiant energy (Garg & Garg, 2017). Few examples of energy harvesting for IoT devices as discussed by (Elahi et al, 2020) are:

- **Wireless Solar Tag (Sol Chip SCC-S433):** It is an ultra-compact, maintenance-free wireless tag based on solar power. It is capable of powering and wireless connecting a large variety of sensors to the cloud. Some of the applications of the wireless solar tag are smart irrigation, precision agriculture, smart cities, smart grid, and smart parking all other similar applications. Being a wireless device the cost and time needed for deploying and maintaining wires to connect sensors is eliminated. It has built-in energy storage and does not require any batteries. Hence there are no maintenance requirements like changing and discarding of batteries.
- **Wibicom's ENVIRO:** It is a circular photovoltaic collector with an antenna of approximately 2 inches in diameter. The device can sense environmental data like temperature, pressure, acceleration, and humidity. One of the interesting features is that the device can operate continuously for two months in the absence of sunlight by using the stored energy. Indirect sunlight can provide a maximum load of 13mw to power sensors and Bluetooth LE.
- **Re Vibe:** It converts vibrations into Ac power to address the issues of predictive maintenance. It uses electromagnetic induction to convert vibration into electricity that can be used to power wireless sensors and

monitoring systems. The two types of harvesters generally known as ModelA and ModelD can power few sensors at a time when connected to a vibration source. By using a constant vibration frequency of 15 to 100 Hz ModelA can produce 150 milliwatts while ModelD can produce 40 milliwatts. These devices can store energy for later use.

- **Tego RFID Tags:** The RFID devices made by Tego use the carrier signal of the RFID reader to rectify it into a DC voltage. This generated power which may be around 4 milliwatts is used to power the processor of the RFID chip. The device adds readable, writeable, and encrypted data to any type of asset. These chips operate in a range of five to ten feet from the reader. The Tego tags with serial interface on their chips are suitable to be connected with the sensors or microprocessors.

Energy Efficient Routing Protocols

Routing plays a vital role in establishing communication among the devices in the networks by deciding the best route among the source and sink nodes. Energy is consumed by the nodes during the processing and transferring of the voluminous IoT data. The nodes are generally energy-constrained and repeated usage of the nodes will fail the nodes. The presence of dead nodes on the routing path can disrupt the whole communication. Along with less energy-consuming technologies and energy harvesting techniques discussed in the previous sections, it is equally important to develop routing protocols that will minimize the energy consumption of the nodes (Kumar Poluru & Naseera, 2017).

The research community has been trying to devise protocols that will enable reliable routing with the security of data, avoid congestion of the networks and minimize the energy consumption of the nodes. The energy consumed during data transfer is proportional to the distance between the sources and sink nodes. One of the methods to reduce the power consumption of a routing path is to select paths with the shortest distance. Some of the routing protocols that can increase the energy efficiency of the IoT networks are bio-inspired routing, fuzzy-based approaches, energy harvesting, and IPV6 for low power and lossy networks (RPL) (Gopika & Panjanathan, 2020). Some of the techniques that can be adopted while developing routing protocols to enhance the energy efficiency of the networks are:

- For a static environment, the nodes can be divided into different regions and a group of nodes can be used based on the route discovery.
- The energy consumption of the low power devices can be managed by configuring the nodes with energy harvesting capabilities and storing the energy to be used whenever the need arises. When the remaining energy of

the nodes is less than the minimum energy required to operate, the energy harvesting process is initiated.

- Sharing of energy among the nodes should be done in which the energy of the high energy gateways will be transferred to the lower energy nodes. By this process no nodes will run out of energy and the stability of the networks will be improved.

- The remaining energy of the nodes and hop count can be considered important parameters while working with power-constrained devices. The number of intermediate nodes should be less and the next-hop neighbor should be selected based on the overall energy of the network. The nodes with less residual energy should be avoided.

- The nearest node should be selected as the next-hop neighbor to reduce energy consumption.

- Using a clustered-based approach for routing can reduce congestion and thus improve the network lifetime.

- The route discovery process can be improved by utilizing bio-inspired algorithms based on swarm intelligence like ant colony optimization (ACO). It uses the foraging behavior of ants to find the shortest path on a given route and thus reduces energy consumption.

- Fog computing approaches can reduce the latency and the energy consumption as they are nearer to the users.

CONNECTIVITY IN IoT

In IoT, digital devices communicate with each other by connecting through the internet. As the IoT devices are connected on a global scale and the number of devices or sensors on the network increase the connectivity issues tend to increase. Data transfer in IoT significantly depends on the connectivity medium. The disruptions arising due to the connectivity standards or other related issues can affect the functioning of the IoT applications. Some of the enabling technologies or the connectivity standards generally use for IoT are Wi-Fi, Bluetooth Low Energy, Li-fi, Zigbee, Thread, Z wave, RFID, and near field communication. One of the reasons that make connectivity challenging for the IoT networks is that the internet used for establishing the connection is a heterogeneous network. The different companies or service providers use different standards and technologies. The proxy servers, firewalls, and cellular networks used by the companies are different most of the time. Some may use fast connectivity while others may use slow connectivity.

Figure 5. IoT connectivity technologies
(Postscapes and Harbor Research 2015)

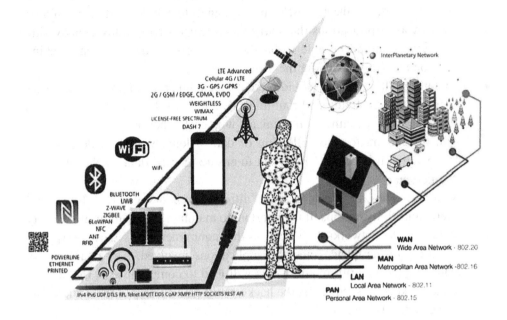

Challenges of Connectivity

Some of the elements that give rise to the connectivity issues in IoT include bandwidth, signaling, interoperability, and power consumption (Samuel, 2016).

- **Interoperability:** The different types of connectivity standards should be chosen carefully to avoid connectivity issues. The organizations involved in providing IoT-based services have to select among the different types of devices and sensors that need to be deployed for an application. These components may run on different standards and protocols. These factors may give rise to connectivity issues in case the devices from different vendors are not compatible with each other or interoperable.
- **Power Consumption:** The components like processors, displays, and computer interfaces have relatively high energy requirements. Connectivity issues may arise if the energy consumption is not managed efficiently as the failure of any one of the nodes or components can disrupt the whole communication path. The power consumption of an IoT network should be minimum to ensure long battery life so that the devices can deliver a great user experience for a long time with small batteries.

- **Bandwidth:** The devices on the network continuously send request/response signals to the servers. Bandwidth requirements are essential to enable the server to handle all the request/response signals to handle the data transfer on the network. Applications that carry video streams require high bandwidth. The servers should be able to handle the enormous data exchange taking place on the network among the devices.

- **Signaling:** The reliability of bidirectional signaling is an important parameter to enable the collecting and routing of data on an IoT network. The devices on the network are either communicate with each other or may communicate with the servers for the collection of data. Along with the side range of connectivity, it is equally important to ensure that the data reaches its exact destination.

- **Security:** The IoT devices from different vendors come with several interconnectivity options. It is important to maintain and monitor the security of data on the networks is difficult as the requirements of security may vary according to the devices and the applications involved. Any deficiency in managing the security requirements will lead to compromise of the devices to the hackers. In such cases, the hackers will gain control of the devices and create connectivity issues.

- **Presence Detection:** Detecting and identifying an IoT device before they shut down before they go offline and when they become online again is termed as presence detection. The nodes have to be monitored continuously as connectivity issues will arise if the nodes are in offline mode. Continuous monitoring will be helpful in the identification and fixing of problems that may arise in the network.

Solutions for Connectivity Issues

Among the numerous available connectivity standards, each has its strength and weakness. It is therefore very difficult to select connectivity standards for an application. Each of the connectivity protocols works on distinct bandwidths and possesses distinct carrying capacities for devices. The connectivity solutions can be classified as wireless and wired. The wired solutions can be used for the applications in which the machine or the thing stays at the same location and there is no need for mobility or when the distance between the sensor and gateway is very less. The wireless solutions are further classified as: (i) Long-range connectivity standards that work in the range of 200 Kms and (ii) Short-range connectivity standards that work in the range of 100 m (Deloitte, 2018)

- **Short Range Solutions:** Some of the frequently used short-range solutions are Bluetooth LE (BLE) and Wi-Fi. BLE generally works in the range of 2.4 to 2,485 GHz while Wi-fi works in the range of 2.4 and 5 GHz (Pradeep et. al, 2016). As already discussed in section 3.1 BLE consumes less power and therefore may contribute to deal with the connectivity issues arising due to the power consumption by the batteries of IoT devices. Using Wi-Fi simplifies the connection of devices on the internet but it increases the power requirements. The power requirement by the Wi-Fi can be reduced by creating a separate Wi-Fi for handling certain services and devices. The security issues with Wi-Fi can be addressed by installing designated IoT networks with extra security layers (Asthana, 2019). Z-Wave is one more short-range connectivity protocol that is energy efficient and transfers data at a frequency of 900 MHz (Lihn & Kim, 2018). It operates at a lower bandwidth as compared to BLE and Wi-Fi. Z-Wave and is energy efficient thereby helping to deal with power consumption and bandwidth issues faced during connectivity of IoT devices.
- **Long Range Solutions:** This category of solutions consists of cellular networks and the Low Power Wide Area Network (LPWAN). Cellular networks are preferred for IoT solutions as they are capable of transferring a high amount of data for long distances at low latency. The recent cellular standards like the LTE-Cat-0, 1, M1, NB1 can provide low power, low throughput wireless technology needed by the IoT applications. The LPWAN can operate on miniature inexpensive batteries as its hardware power consumption is very low. It is suitable to be used for urban areas in an operating range of 10 km. It is not just a connectivity standard but is an umbrella term that encompasses several implementations and protocols that work on some kind of connectivity features.

The organizations working with IoT have to collect a large amount of data. The data has to be captured and then downloaded to a local server in the local office. The problems of bandwidth can be solved by using broadband networks for applications that involve video surveillance, photo collection, and video conferencing. It is not easy to implement a uniform standard of interoperability for all the devices. The service providers face difficulties in connecting the devices of different standards to enable data communication among them. This problem can be solved to some extent by testing the compatibility of the devices and associate with vendors that can provide alternate solutions for integrating their products and services. Many of the IoT devices do not have any security provisions. The security standards should be checked by the service providers before installing the devices. Implementing network segmentation, changing default passwords, regular updates of software and firmware are some of the measures that can be adopted to overcome connectivity issues developed due to lack of security measures (Shacklett, 2021).

SECURITY IN IoT

Security is one of the most significant challenges faced by the IoT industry that needs immediate attention to ensure secure and reliable connectivity to its users. The increase in the number of IoT devices increases the attack surface as any of the devices could be a possible target. Many IoT devices lack security standards to prevent them from hacking. The service providers and the organizations are aware of the significance of protecting the IoT devices. But due to the high investments and lack of knowledge among the users, security is kept at the bottom of the priority list. Lack of security can be dangerous and it may also lead to a lack of trust about the IoT among its users.

Security Challenges in IoT

In the absence of authenticity, confidentiality, and privacy it is not possible to encourage large-scale usage of IoT among the users. The three major security challenges as shown in fig 5 are data confidentiality, privacy, and trust (Miorandi et.al, 2012).

- **Data Confidentiality:** This is mainly applicable in the business context where data represents an asset. The data has to be kept secure to conserve competitiveness and market value. This involves taking care of two important issues: first is defining an access control mechanism and the second is defining an authentication process for devices. Data confidentiality remains a primary concern for most IoT applications as its data is related to the physical realm. If we consider an example of environmental monitoring application that provides early warnings of natural calamities, it becomes extremely important that the data is accessed only by the relevant civil protection bodies. Leaking of such information among the common people may create panic and create hindrance in the implementation of the risk management strategies by the civic bodies.

The traditional approaches used for securing data over the internet cannot be directly applied for IoT mainly due to the enormous data generation by the IoT devices. The second challenge in this regard is to control the access of data by the relevant users with the flexibility to change the access rights at run time in case of dynamic data streams. Identity management of the devices and users is another aspect

Figure 6. Security challenges in IoT
(Miorandi et.al, 2012)

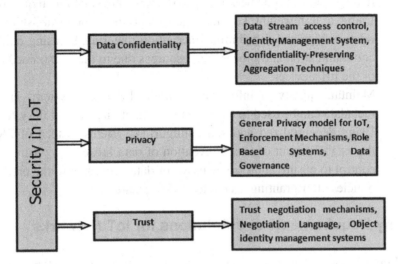

that has to be considered while solving the confidentiality issues especially for the IoT scenarios that involve fusion of the digital and physical world. Unauthorized access can be avoided by combining access control mechanisms with suitable data protection techniques.

- **Privacy:** It involves safeguarding the personal information of the users and controlling the usage of the information by authorized agencies. The wireless channels for the exchange of data and the remote accessing capabilities increase the challenges of ensuring privacy for IoT applications. Some of the open research challenges for the privacy-preserving mechanisms in IoT scenarios are:
  ○ Defining a general model for privacy
  ○ Developing novel enforcement methods that can support the scalability and heterogeneity issues
  ○ Defining privacy policies to identify smart objects and specify the conditions to access the sensitive data.
- **Trust:** A trust framework is required to enable the users of the IoT applications to have confidence regarding the reliability of the information and services that are exchanged on the IoT networks. The trust framework should be able to handle the humans as well as the machines by developing trust among the humans and by being strong enough to be used by the machines without denial of service. Some of the challenges in the development of a trust-based framework are (Vermasan & Freiss, 2014):

- ○ Using lightweight Public Key Infrastructure (PKI) as a basis for trust management and enable solutions to address the scalability requirements.
- ○ Using a lightweight key management system for establishing trust relationships and distributing encryption materials by using minimum communication and processing resources due to the constrained nature of the IoT devices.
- ○ Maintain quality of information for the IoT-based systems in which metadata can be used to evaluate the authenticity of the IoT system.
- ○ Using decentralized and self-configuring systems instead of PKI for the establishment of trust Prevention of data breaches by using access control to ensure appropriate usage of data according to the predefined policies after granting access to the information.

Security Requirements and Solutions for IoT Networks

The IoT devices on the network are an easy target for intrusion due to their resource-constrained nature. The huge number of devices and proportionally huge data increase the complexity of securing the IoT networks. Some of the securities requirements of the IoT networks are privacy, confidentiality, secure routing, robust and resilient management, and attack detection (Hameed et al. 2019).

- **Privacy:** One of the challenges while assuring privacy is profiling and tracking that is caused due to the association of an identity of a certain individual. Another threat is localization as the systems try to record the location of humans by using the information of time and space. It is essential to design protocols that would are able the activities like profiling and localization. The third challenge is ensuring the secure transfer of data without revealing any information to unauthorized users. These challenges can be encountered using the following measures:
  - ○ Using a comprehensive privacy-preserving framework that will use a generic lightweight cryptographic privacy-preserving algorithm to ensure confidential exchange of data and preserve the origin of data.
  - ○ Using context-aware privacy policies to preserve the privacy of data streams in IoT by utilizing the data management policies and control mechanisms for dynamic data access.
  - ○ Using game theory to analyze location privacy and improving the existing privacy-preserving protocols.
  - ○ Using software-defined networking (SDN) for network virtualization to preserve the privacy of the large amount of data being transferred or exchanged on the IoT networks. SDN helps the network operators

to implement privacy over the whole network by centralization of the routing and forward functionality.

- **Lightweight Cryptographic Framework:** Taking into consideration the hardware constraints of the IoT devices there is a need to develop lightweight cryptographic solutions. These solutions should consume fewer resources without compromising the preferred security levels. The designing of the lightweight cryptographic framework should consider the abilities and lifecycle of an IoT device. Some more factors that need to be considered are scalable architectures that allow the security domains to be used for small scale to large scale IoT deployments, including aspects of trusted third party and type of protocols used and using security at all the levels of IoT architecture (sensing, network, application) due to varied security prerequisites and communication patterns at each level. Efficient holistic frameworks should be developed that will be able to use lightweight cryptographic frameworks. The existing solutions were designed for machines with sufficient energy resources, computing capabilities, and storage space. The cryptographic solutions used for the internet cannot be directly used for IoT due to the heterogeneous and constrained nature of the IoT devices. Central monitoring and reconfiguration of the networks by using SDNs open new prospects for applying lightweight cryptographic frameworks so that lightweight security solutions can be used for the SDN controller.

- **Secure Routing:** The data packets are transmitted from node to node on the route till they reach their destination. While traveling through the route the data packets may be subjected to different types of attacks from the malicious nodes resulting in unnecessary delays or loss of information. To overcome this, routing protocols should be able to establish a secure route for the data packets and use lightweight computations methods for routing data in the low-powered IoT networks. Most of the IoT networks are self-organizing that work without human intervention. It, therefore, becomes easy for outsiders to introduce malicious nodes in the network. The routing protocols should be able to detect and isolate the malicious nodes from the network. In addition to this, there is a need to design secure routing protocols that would be self-stabilizing so that the networks would be able to recover immediately without human involvement. Location privacy of the source and destination nodes is another challenge that should be taken care of while developing security protocols. The security protocols should be able to preserve location privacy. To fulfill the above-mentioned requirements of a secure routing protocol it is essential to consider the IoT network performance so that the resource limitations are known and lightweight mechanisms can be developed to mitigate the attacks. Apart from this, there should be a provision to track the situation of the whole network and use routing control policies that can be altered rapidly to prevent the effect of security attacks.

- **Robust and Resilient Management:** The disturbance caused in the IoT applications due to system failures will interrupt everyday activities and may sometimes prove to be hazardous to the lives of its users. The challenges observed while implementing robust and resilient management are:
  - ○ Developing novel algorithms that are inherently tolerant to the malicious attacks in the networks. Developing methods that will be able to detect the attacks before it spreads across the network and damage it.
  - ○ Disruption of the IoT services for long periods can be life-threatening for people in case of applications like disaster management and health. The resource management framework should be able to detect attacks and at the same time should be able to provide recovery from the attacks. Some possible solutions for ensuring robustness in IoT networks are:
    - ▪ Considering resource constraint nature of the IoT devices while developing protocols for network management.
    - ▪ Controlling the failures of the IoT networks by centralization of network view. As the decision making will be done by the controller, central detection of faults can be done. After detecting faults the controller should be able to decide to reroute the network traffic by using another server or route.
    - ▪ The frameworks should be able to perform detection of faults at appropriate times to tackle the situation by implementing the best possible solutions.
- **Attack Detection:** Denial of Service (DoS) tends to disrupt the normal operations of IoT. The detection of DoS attacks is difficult as they are launched by multiple attackers at the same time. Similarly detecting and addressing the risks of insider attacks is equally challenging as they are launched by the use of unknown devices. Some of the risks involved with insider attacks are the leaking of confidential data and disturbing the operation of IoT networks by launching DoS attacks. Most of the available DoS detection solutions are suitable for the traditional internet. They cannot be directly applied to IoT networks due to the resource-constrained nature of the IoT devices. After detection, developing its mitigation measures becomes challenging due to the requirement of lightweight and energy-efficient solutions. Some of the possible measures to handle this situation are:
  - ○ Implementing novel lightweight solutions that would be appropriate for the resource-constrained nature of the IoT devices.

○   By using SDN to monitor the flow of activities on the IoT networks, it is possible to develop solutions for the detection of attacks and malicious activities.

○   Integrating gateways with the SDNs to efficiently detect and mitigate the effects of DoS attacks.

Table 1. Security requirements and solutions for the IoT Networks

Security Requirements	Challenges	Solutions/Future Directions
Privacy	• Profiling and tracking localization • Secure data transmission	• Comprehensive privacy-preserving frameworks • Context-Aware privacy policies • Game theory-based privacy-preserving incentives • Network virtualizations by using SDN
Lightweight Cryptographic Framework	• Lightweight primitives • Consume low resources	• Efficient holistic frameworks • Utilization of SDNs for lightweight security provisioning
Secure Routing	• Secure route establishment • Isolation of malicious nodes • Self-stabilization of the security protocol • Preservation of location privacy	• IoT network performance-focused routing protocol design • Effective and fine-grained control over routing activities leveraging SDN
Robust and Resilient Management	• Attack tolerance • Early detection of attacks • Quick recovery from failures	• Centralized management frameworks
Attack Detection	• Resource-efficient DoS attack detection • Resource-efficient countermeasures • Resource-efficient insider attack detection	• Lightweight solution for a resource-constrained device • Centralized SDN detection and mitigation algorithms

(Hameed et al. 2019)

ADDRESSING AND IDENTIFICATION

Addressing and identification of the devices remain a challenge as the number of devices on the network is continuously increasing. The millions of smart devices connected to the internet have to be identified by a unique address. The unique identification of each device will require a large addressing space for managing and controlling them remotely through the internet (Farhan et.al 2018). Reliability, scalability, distinctiveness, and persistence are some of the requirements of creating a unique address for smart devices (J.Gubbi et.al 2013). A suitable and unique address will help the devices to communicate over the internet. The devices have to identify their neighboring devices and be aware of the services provided by them. Scalability has to be considered as an important feature while generating unique addresses for smart devices because the number of devices on the network may increase in the future (Sethi and Sarangi, 2017). Each device that is introduced in the network has to be installed and monitored for its functioning to diagnose and update its problems. The management of these devices becomes difficult as many of them are installed at remote locations and are inaccessible.

Addressing Schemes in IoT

Internet Protocol Version 4 (IPV4) enabled devices to connect to the internet using a 32-bit address. This addressing capacity has been exhausted as it could provide only 4.3 billion IP addresses. The limited address space and functionality of IPv4 were unable to handle the rapidly increasing number of devices. Another generation of addressing scheme available is IPV6 128 bit addressing that allows 3.4×10^{38} addresses that are equal to 340 trillion trillion trillion IP addresses (Bajrami, 2019).

There are several technical challenges while assigning a unique address to the devices using IPv6 in the IoT environment. Some of these challenges include (Kumar & Tomar, 2018):

1. The constrained IoT nodes will not be able to generate unique addresses due to limited processing capabilities and the absence of input and output interface.
2. Regeneration of address to the IoT devices without affecting the ongoing operations in situations when the devices move from one network to the other.
3. There are multi-homing challenges that arise when the devices change between IPv6 and non-IPv6 networks.
4. The unique ID of the nodes working as bridge or proxy for the non-IP network may keep on changing. As these nodes do not have permanent IPv6 addresses tracking of non-IP nodes is difficult.
5. The IPV6 addresses may not be unique locally or globally. Duplicate address detection sometimes becomes a challenge.
6. Generation of unique address for the IoT nodes using IPV6 addressing mechanism.
7. Maintaining minimum energy consumption of the addressing schemes.
8. Avoiding communication overhead generate due to the generation of some requests and replies during the address generation.

The researchers have studied the challenges involved with IPv6 and proposed some methods to handle them. A lightweight 6HOP addressing scheme (Aljosha et.al, 2017) and lightweight resource addressing (Bingqing & Zhixin, 2015) are some of the techniques used to improve the performance of IPv6. The IPv6 addressing mechanisms generally produce static or predictable addresses that increase the chance of attacks. In 6HOP addressing technique, the addresses are generated in a predetermined manner by the communication but they appear to be random to the third parties. The IoT servers that use the dynamic address format of the 6HOP change their addresses and ports at regular time intervals to protect the networks against attacks. The lightweight addressing approach enables ample use of IPv6 addresses space, offers superior security characteristics, and makes deployment easier. The lightweight resource addressing approach uses two types of addressing modes to solve the problem of heterogeneous encoding. This model uses direct addressing mode for active nodes and indirect addressing modes for the passive nodes. It uses a virtual domain to deal with the problem of heterogeneous encoding. It provides suitable interconnection between the WSNs and the IPv6 using the 6LoWPAN protocol.

Identification in IoT

Identifiers in IoT are defined as patterns that enable the unique identification of a single entity known as an instance identifier or a class of identity known as a type identifier within a specific context. A survey by the Alliance of Internet of Things Innovation provides a high-level discussion about the different types of identifiers and their requirements (AIOTI, 2018). The different types of identifiers are:

- Thing identifiers that identify the entities that can interact with each other. The entities include physical objects like humans, animals, plants, or machines and digital data like files and datasets.
- Application & Service identifiers that identify software applications and services.
- Communication identifiers to identify the source and destination endpoints in communication.
- User identifier to identify the IoT applications and services.
- Data identifiers that identify different data types and specific data instances.
- Location identifiers that identify locations within a geographical area.
- Protocol identifiers that identify communication protocols required for specific communication exchange.

As there are different types of identifiers for IoT intended for a specific purpose similarly these identifiers have a long list of requirements. Some of the requirements include (i) Uniqueness of the identifier which can be assigned either globally or locally according to the requirements of a specific application (ii) Privacy and personal

Figure 7. IoT identifiers according to AIOTI high-level architecture
(AIOTI, 2018)

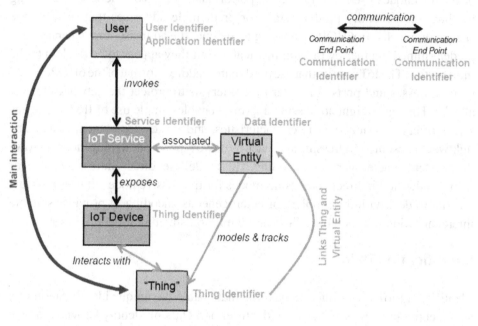

data protection for the applications that involve humans and entities related to them (iii) Security of data to avoid its tampering during transfer or usage. (iv) Identifier patterns that are suitable to be used in constrained environments (v) Traceability to identify the origin of the identified entity and proofing its authenticity (vi) Scalability of the identifier pattern and its lifecycle management. (vii) Interoperability of the identifier schemes for the applications from different domains, regions, and industries. (viii) Persistency and reuse of identifiers during the lifetime of an entity. (ix) Allocation of own set of identifiers by organizations so that there are no conflicts with other organizations. (x) Registration of identifiers on a global database to store the information of the identified entities that will be available later either online or offline according to the applications.

The study of various identifiers and their requirements helps the developers to select identifiers and the requirements according to specific applications. It also helps in determining relevant identification schemes (ISS) to satisfy their needs. There are various ISs in use for years but most of them are domain-specific. The unavailability of ISs that would fit all the needs, increase the challenge of inter-networking among the IoT platforms. The identification schemes should be designed considering that an IoT system should be able to interact with other systems. Communication among the IoT applications running on different platforms is possible by developing a unique identification system. Though universal ISs are available they are no single universal IS that can be used by all the platforms of the IoT. Achieving interoperability becomes difficult to achieve as the different platforms use different ISs (Liu et.al, 2014). The ISS for the IoT is based on the concept of fulfilling conditions of management, maintenance, and mapping discovery of identifiers and recovering identifier information. A common identification mechanism for IoT is difficult to develop due to many reasons but understanding the working of various ISs can help develop a unified scheme (Aftab et. al, 2019). Some of the universal ISs are Object Identifier (OID), Universal Unique Identifier (UUID), and Electronic Product Code (EPC).

- **OID:** The OID is an identification scheme jointly developed by ISO/ IEC (International Organization for Standardization/ International Electrochemical Commission) and ITU-T (International Telecommunication Union Telecommunication Standardization Sector). Its hierarchical tree structure provides the flexibility to extend the layers and length of the identifiers. The tree-like structure enables the assignment of a unique number to the objects for preventing duplication. An OID is capable of identifying any physical or virtual objects and connecting them to the global information and communication infrastructures. It can accommodate short-range communication technologies. It provides a harmonized way to for integrating identifiers of the devices that require communication capabilities. It also identifies the devices in which communication capabilities are unnecessary. It can be used for tag-based identification to identify a physical object by using the information stored in servers. The OIDs are suitable for working with different kinds of networking technologies and hence help in solving the interoperability issues. It can identify devices from different layers by using the identifier schemes. It is therefore able to fulfill the prerequisites of mapping identifiers to objects from different layers and integrating them effortlessly (ITU-T, 2017).
- **UUID***:* A UUID also known as GUID (Globally Unique Identifier) is a 128-bit number that enables a unique identification of an object or entity on the internet. UUID can be used for the identification of users by relating to their preference cards. The randomly generated UUID is validated for its existence on the server. In case the validation is positive, UUID is associated with the preference card of the user. In case of negative validation, a new random UUID is generated and the validation process begins again. The user can locally store or export the UUID created for his email and use it for more than one device in the system. Only the UUID and preference cards are transferred to the server so there is no possibility of identifying the user thus reducing the possibility of attack vectors. As the UUID does not require centralized

authority, it allows the generation of demand to be fully automated. The UUIDs are unique and the chances of duplications are negligible. This feature of UUID makes it an excellent option for using it as a universal IS. It can be used for multiple objectives like tagging objects with an extremely short lifetime or identification of persistent objects across the networks. (Leach & Mealling, 2005).

- **EPC:** The EPC was designed by Auto-ID from the Massachusetts Institute of Technology as an eventual successor to the bar code. It was created to share data in real-time by finding a unique identifier and using RFID through internet infrastructure and platform. The primary aim of EPC is to support the use of RFID and expand it to the worldwide network for the creation of a smart industry for standard global for EPC network (Hallaj Asghar, 2015). The unique identifier assigned by the EPC global architecture framework has the following characteristics:
  - All the entities carry a serialized unique number.
  - An EPC for each entity is globally unique regardless of its type.
  - Compatible and interoperable with the existing naming systems.
  - The existing naming systems are allowed to be incorporated into the EPC to achieve universality.
  - Additional naming systems are allowed to be incorporated without affecting the existing systems.
  - The EPCs are designed to support the decentralized assignment of the new EPCs without the probability of collisions.
  - It is a lightweight naming system as the other information associated with the EPC bearing devices is not encoded into EPC but associated through other means (GS1,2014).

Some of the solutions for the IoT identifications as given by the European Research Cluster on the Internet of Things (IERC, 2014) are:

- One of the solutions is using a system architecture that includes a resolver, name server, and information server. Resolver is an IoT ID query software client that works as a library procedure so that it can be called by any type of application. The name servers store the different kinds of resource records and respond to queries against these records. The information server stores detailed information of the resources and enables capturing of events and accepting their queries.
- The heterogeneity issue of the identification schemes can be solved by using a naming mechanism with two-stage identification. The first stage is the standard identifier (SID) that uses a unique identifier for each naming

scheme. The second stage called the resource identifier (RID) uses unique identification for each object connected to the internet infrastructure.

- Integrated naming and addressing solutions based on IPv6 can be used to provide the mapping of legacy IoT IDs to IPv6. Using this method eliminates the need to reveal any information of the IoT end-point configurations by the IPv6 addresses.

- Semantic interoperability infrastructure can be used for naming and discovery for the IoT applications like smart cities that include many integrated applications. One of the examples of this kind of infrastructure is a cloud-based directory module in which the IoT resource instances are registered by using a universal resource identifier (URI). The cloud discoverer module working over the directory allows finding the resources by location, by location and type, and by using URI.

CONCLUSION

In this chapter, we have studied some of the challenges faced by the IoT networks like data management, security, connectivity, energy consumption, and addressing of the IoT devices. The chapter also suggests some possible solutions to overcome these challenges. It may not be possible to overcome all the challenges at a time. But to expand the usage of IoT, it is essential to keep on updating oneself with the probable challenges, their causes, and methods to overcome them.

REFERENCES

Abu-Elkheir, M., Hayajneh, M., & Ali, N. (2013). Data Management for the Internet of Things: Design Primitives and Solution. *Sensors (Basel)*, *13*(11), 15582–15612. doi:10.3390131115582 PMID:24240599

Aftab, Gilani, Lee, Nkenyereye, Jeong, & Song. (2019). Analysis Of Identifiers On IoT Platforms. *Digital Communications and Networks*. doi:10.1016/j.dcan.2019.05.003

Aljosha, J., Johanna, U., Georg, M., Voyiatzis, A. G., & Edgar, W. (2017). Lightweight Address Hopping for Defending the IPv6 IoT. *Proceedings of ARES '17*.

Alliance For Internet of Things Innovation. (2018). *Identifiers in the Internet of Things (IoT)*. https://euagenda.eu/upload/publications/identifiers-in-internet-of-things-iot.pdf

An, Minh, & Kim. (2018). A Study of the Z-Wave Protocol: Implementing Your Own Smart Home Gateway. *2018 3rd International Conference on Computer and Communication Systems (ICCCS)*. 10.1109/CCOMS.2018.8463281

Asghar, M. H. (2015). RFID and EPC as key technology on Internet of Things (IoT). *International Journal of Clothing Science and Technology*, 6(1).

Asthana, R. (2019). *Solving Common IoT Challenges*. https://dzone.com/articles/the-most-effective-method-to-overcome-4-common-iot

Bajrami, V. (2019). *What You Need to Know About IPv6*. https://www.redhat.com/sysadmin/what-you-need-know-about-ipv6

Bonomi, F., Milito, R., Zhu, J., & Addepalli, S. (2012). Fog computing and its role in the internet of things. *Proceedings of the First Edition of the MCC Workshop on Mobile Cloud Computing - MCC '12*. 10.1145/2342509.2342513

Cetinkaya, O., & Akan, O. B. (2015). A DASH7-based power metering system. *12th Annual IEEE Consumer Communications and Networking Conference*. 10.1109/CCNC.2015.7158010

Cha, H.-J., Yang, H.-K., & Song, Y.-J. (2018). A Study on the Design of Fog Computing Architecture Using Sensor Networks. *Sensors (Basel)*, 18(11), 3633. doi:10.339018113633 PMID:30373132

Chen, X.-Y., & Jin, Z.-G. (2012). Research on Key Technology and Applications for Internet of Things. *Physics Procedia*, 33, 561–566. doi:10.1016/j.phpro.2012.05.104

Deloitte. (2018). *The Future of Connectivity in IoT Deployments*. Author.

Elahi, H., Munir, K., Eugeni, M., Atek, S., & Gaudenzi, P. (2020). Energy Harvesting towards Self-Powered IoT Devices. *Energies*, 13(21), 5528. doi:10.3390/en13215528

Fafoutis, X., Tsimbalo, E., Zhao, W., Chen, H., Mellios, E., Harwin, W., Piechocki, R., & Craddock, I. (2016). BLE or IEEE 802.15.4: Which Home IoT Communication Solution is more Energy-Efficient? *EAI Endorsed Transactions on Internet of Things*, 2(5), 1–8. doi:10.4108/eai.1-12-2016.151713

Farhan, L., Kharel, R., Kaiwartya, O., Quiroz-Castellanos, M., & Alissa, A. (2018). A concise Review on Internet of Things (IoT) – Problems, Challenges and Opportunities. *11th International Symposium on Communication Systems, Networks, and Digital Signal Processing (CSNDSP)*. 10.1109/CSNDSP.2018.8471762

GS1. (2014). *The GS1 EPCglobal Architecture Framework*. https://www.gs1.org/sites/default/files/docs/epc/architecture_1_6-framework 20140414.pdf

Garg, N., & Garg, R. (2017). Energy harvesting in IoT devices: A survey. *2017 International Conference on Intelligent Sustainable Systems (ICISS)*, 10.1109/ ISS1.2017.8389371

Gopika, D., & Panjanathan, R. (2020). Energy-efficient routing protocols for WSN based IoT applications: A review. *Materials Today: Proceedings*. Advance online publication. doi:10.1016/j.matpr.2020.10.137

Gubbi, J., Buyya, R., Marusic, S., & Palaniswami, M. (2013). Internet of Things (IoT): A vision, architectural elements, and future directions. *Future Generation Computer Systems*, *29*(7), 1645–1660. doi:10.1016/j.future.2013.01.010

Hameed, Khan, & Hameed. (2019). Understanding Security Requirements and Challenges in Internet of Things (IoT): A Review. *Journal of Computer Networks and Communications*.

IERC-European Research Cluster on the Internet of Things. (2014). *Internet of Things*. Joint White Paper on Internet-of-Things Identification.

Iorga, M., Feldman, L., Barton, R., Martin, M. J., Goren, N., & Mahmoudi, C. (2018). *Fog Computing Conceptual Model*. doi:10.6028/NIST.SP.500-325

ITU-T. (2017). *ITU-T X6.660-Supplement on Guidelines for Using Object Identifiers for Internet of Things*. Author.

Kirtania, S. G., Younes, B. A., Hossain, A. R., Karacolak, T., & Sekhar, P. K. (2021). CPW-Fed Flexible Ultra-Wideband Antenna for IoT Applications. *Micromachines*, *12*(4), 453. doi:10.3390/mi12040453 PMID:33920716

Kumar, G., & Tomar, P. (2018). A Survey of IPv6 Addressing Schemes for Internet of Things. *International Journal of Hyperconnectivity and the Internet of Things*, *2*(2), 43–57. doi:10.4018/IJHIoT.2018070104

Laszewski, T., & Nauduri, P. (2012). *Migrating to the Cloud: Client/Server Migrations to the Oracle Cloud*. Elsevier. doi:10.1016/B978-1-59749-647-6.00001-6

Leach & Mealling. (2005). *A Universally Unique IDentifier (UUID) URN Namespace*. https://www.researchgate.net/publication/215758035_A_Universally_Unique_ IDentifier_UUID_URN_Namespace

Liu, C. H., Yang, B., & Liu, T. (2014). Efficient Naming, Addressing and Profile Services In Internet-of-Things Sensory Environments. *Ad Hoc Networks*, *18*, 18. doi:10.1016/j.adhoc.2013.02.008

Luo, B., & Sun, Z. (2015). Research on the Model of a Lightweight Resource Addressing. *Chinese Journal of Electronics*, *24*(4), 832–836. doi:10.1049/cje.2015.10.028

Mell & Grance. (2009). *A NIST Notional Definition of Cloud Computing, version 15*. NIST.

Miorandi, D., Sicari, S., De Pellegrini, F., & Chlamtac, I. (2012). Internet of Things: Vision, Applications and Research Challenges. *Ad Hoc Networks*, *10*(7), 1497–1516. doi:10.1016/j.adhoc.2012.02.016

Olive, C. (2012). *White Paper: Cloud Computing Characteristics are Key*. https://www.gpstrategies.com/wpcontent/uploads/2016/04/wpCloudCharacteristics.pdf

Poluru, R. K., & Naseera, S. (2017). A Literature Review on Routing Strategy in the Internet of Things. *Journal of Engineering Science and Technology Review*, *10*(5), 50–60. doi:10.25103/jestr.105.06

Postscapes and Harbor Research. (2015). *What Exactly is the Internet of Things*. https://www.visualistan.com/2015/09/what-exactly-is-internet-of-things.html

Pradeep, Kousalya, Suresh, & Edwin. (2016). IoT And Its Connectivity Challenges In Smart Home. *International Research Journal of Engineering and Technology*, *3*(12).

Puthal, Sahoo, Mishra, & Swain. (2015). Cloud Computing Features, Issues and Challenges: A Big Picture. *International Conference on Computational Intelligence & Networks*.

Rountree, D., & Castrillo, I. (2014). *Cloud Deployment Models, The Basics of Cloud Computing*. Elsevier.

Samuel, S. S. I. (2016). A Review Of Connectivity Challenges In Iot-Smart Home. *2016 3rd MEC International Conference on Big Data and Smart City (ICBDSC)*. 10.1109/ICBDSC.2016.7460395

Sasikala, P. (2013). Research Challenges And Potential Green Technological Applications In Cloud Computing. *International Journal of Cloud Computing*, *2*(1), 1–19. doi:10.1504/IJCC.2013.050953

Sethi, P., & Sarangi, S. R. (2017). Internet of Things: Architectures, Protocols, and Applications. *Journal of Electrical and Computer Engineering*, *2017*, 9324035. doi:10.1155/2017/9324035

Shacklett, M. (2021). *4 IoT Connectivity Challenges and Strategies to Tackle Them.* https://internetofthingsagenda.techtarget.com/feature/4-IoT-connectivity-challenges-and-strategies-to-tackle-them

Sharma, & Vijay, & Shukla. (2020). Ultra-Wideband Technology: Standards, Characteristics, Applications. *Helix, 10*(4), 59–65. doi:10.29042/2020-10-4-59-65

Subrahmanyam, V., Zubair, M. A., Kumar, A., & Rajalakshmi, P. (2018). A Low Power Minimal Error IEEE 802.15.4 Transceiver for Heart Monitoring in IoT Applications. *Wireless Personal Communications, 100*(2), 611–629. doi:10.100711277-018-5255-y

Varghese, B., Reano, C., & Silla, F. (2018). Accelerator Virtualization in Fog Computing: Moving from the Cloud to the Edge. *IEEE Cloud Comput, 5*(6), 28–37. doi:10.1109/MCC.2018.064181118

Vermasan, O., & Friess, P. (2014). *Internet of Things- From Research and Innovation to Market, Deployment.* River Publishers.

Wadhwa, H., & Aron, R. (2018). Fog Computing with the Integration of Internet of Things: Architecture, Applications, and Future Directions. IEEE Intl Conf on Parallel & Distributed Processing with Applications, Ubiquitous Computing & Communications, Big Data & Cloud Computing, Social Computing & Networking, Sustainable Computing & Communications (ISPA/IUCC/BDCloud/SocialCom/SustainCom), 987-994.

Chapter 4
IoT Sensors:
Interference Management
Through Power Control

Weston Mwashita
Vaal University of Technology, South Africa

Marcel Ohanga Odhiambo
Mangosuthu University of Technology, South Africa

ABSTRACT

This research presents a power control method for proximity services (ProSe)-enabled sensors to enable the sensors to be seamlessly incorporated into 5G mobile networks. 5G networks are expected to create an enabling environment for 21st-century advancements such as the internet of things (IoT). Sensors are crucial in the internet of things. A power control strategy involving two power control mechanisms is proposed in this research work: an open loop power control (OLPC) mechanism that can be used by a ProSe-enabled sensor to establish communication with a base station (BS) and a closed loop power control (CLPC) mechanism that can be used by a BS to establish the transmit power levels for devices involved in a device to device (D2D) communication. Several studies have proposed power control strategies to mitigate interference in D2D-enabled mobile networks, but none has attempted to address the interference caused by ProSe-enabled sensors communicating with smart phones and 5G BSs.

INTRODUCTION

In a survey conducted by Ericsson in 2017, 92% of executives in charge of at least

DOI: 10.4018/978-1-7998-9312-7.ch004

Figure 1. Adding a ProSe-enabled sensor to the architecture. Adapted from 3rd Generation Partnership Project (3GPP) Organisational Partners (2017).
Source: 3GPP, 2017

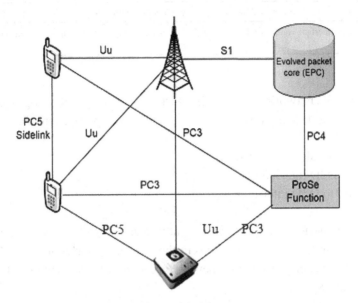

ProSe-enabled sensor

100 of the world's leading telecommunication providers are in agreement that 5G paves the way for a significant number of future technologies. The 5G infrastructure largely serves as a basis for heterogeneous wireless networks, allowing for seamless connectivity that stimulates the growth of smart cities around the world. On the other hand, substantial advancements in hardware manufacturing techniques (Bhushan & Sahoo, 2018) have led to the development of small battery-powered sensors that can connect and communicate with some 5G elements in a D2D manner. These sensors, according to Rathee, Ahuja, and Nayyar (2019), can expand interconnection thanks to IoT. Figure 1 shows ProSe-enabled devices that have ProSe applications running on them.

As seen in Figure 1, a few new interfaces have been developed, the most significant of which are PC3 and PC5 (3GPP 2017:13). Table 1 summarises the functions of these interfaces.

ProSe-enabled devices can communicate with a ProSe Function by using PC3 to collect information for network-related tasks, as shown in Figure 1. ProSe-enabled devices can use PC3 to request permission from a BS to participate in a D2D session. However, this direct connection comes at a cost to overall 5G QoS. When external devices, such as ProSe sensors (3GPP, 2014), communicate directly within

Table 1. Functions of PC3 and PC5

Interface	Function
PC3	This is the interface used by UEs to connect to a ProSe Function
PC5	This interface can be used by UEs or devices like sensors to connect directly to each other in a D2D

an underlying 5G network, the following sorts of interference develop, according to Mach, Becvar, and Vanek (2015). The following types of interference arise:

- D2D to cellular interference.
- D2D to D2D interference.
- Cellular to D2D interference.

The above interferences pose significant threats to cellular networks, prompting numerous researchers to seek solutions to the resulting interference. The following interference management schemes have been identified, according to Mwashita & Odhiambo (2018):

- Resource allocation interference management schemes.
- Power control interference management schemes.
- Retransmission interference management schemes.

Contribution

This study examines transmit power optimisation and provides solutions for usage with ProSe-enabled sensors that need to connect to 5G mobile networks. The study looks at OLPC and CLCP power control techniques that take advantage of the current circumstances of D2D channels. The research team developed an algorithm to properly deal with the interference that ProSe-enabled sensors cause in 5G networks. Extensive simulations revealed that allowing ProSe-enabled sensors to connect directly with neighbouring UEs can result in a 3.6% drop in total user throughput. Both network consumers and network providers find 3.6% to be normal and acceptable. According to Ramasamy (2017:45), a reduction of up to 5% is acceptable to both users and network providers, and this is the value that is frequently captured in service level agreements. The suggested approach can successfully handle interference from both ProSe-enabled sensors and interference from the cellular network to the sensors, which has never been addressed before.

The power control interference management strategies for 3G to 5G mobile systems are described in Section II. Section III summarises the work of other researcher's comparable works. The proposed power control strategy is detailed in Section IV. In Section V, the proposed scheme is simulated and analysed. Section VI concludes the chapter and provides some closing observations.

POWER CONTROL INTERFERENCE MANAGEMENT SCHEMES

When using ProSe-enabled sensors in a cellular network, cellular user equipment (CUE) traffic must take precedence over ProSe-enabled sensor traffic. For mobile wireless technologies, power regulation has always been a crucial design component (Cho, Choi & You 2013). Fast uplink power control has been widely utilized in code division multiple access (CDMA) systems to address the "near-far" problem, which occurs when signals from CUEs closest to a BS obliterate signals from cell-edge CUEs. This occurs when all CUEs broadcast to the BS with the same transmit power level. Fast power control is no longer necessary due to the adoption of orthogonal frequency-division multiple access (OFDMA) and adaptive modulation algorithms. In order to deal with intra-cell interference, OFDMA takes advantage of orthogonality per-cell resources. To deal with shadowing and pathloss, however, power control in the form of gradual power control is still required. There are two types of power control techniques in implemented in 5G, similar to techniques in 3G and 4G: OLPC and CLPC. Figures 2 and 3 depict these two power control systems.

RELATED WORK

Several researchers have investigated power control strategies to decrease interference caused by D2D communications in cellular networks (Song, Niyato, Han, & Hossain, 2015; D. Feng, Lu, Yuan-Wu, Li & G. Feng, 2013; Chen, Liu, H. Li, X. Li, & S. Li, 2016). Fodor, Penda, Belleschi, Johansson, and Abrardo (2013) investigated how power control strategies function in D2D-enabled cellular networks in terms of interference reduction. The researchers looked at power control strategies that used a fixed signal-to-interference-plus-noise-ratio (SINR) objective, fixed transmit power, and open and closed loops. The researchers asserted that if the relevant parameters are appropriately tuned, LTE-based power control systems can achieve near-optimal performance.

To mitigate for large-scale pathloss effects, Abdallah, Mansour, and Chehab (2017) suggested a distributed power control approach that uses distance dependent characteristics existing between the BS and the D2D connection. D2D transmitters

Figure 2. Open loop power control
Source: RF Wireless World, 2018

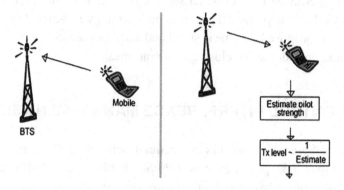

can choose their transmit strengths based on the current channel circumstances under the proposed scheme. Without consulting the BS, the D2D pairs decide on their transmit powers on their own. The technique captures the randomness of BS-D2D distances perfectly. The proposed strategy, on the other hand, did not account for the mobility of the D2D pairings and CUEs. This is especially important in mobile networks because users rarely remain stationary when making or receiving calls. Although ProSe-enabled sensors are normally fixed, the CUEs with which they must communicate are usually mobile. Using this proposed technique with devices like ProSe-enabled sensors could be troublesome because the network QoS is likely to be violated because the BS is not included in the process.

According to Safdar, Ur-Rehman, Muhammad, Imran, and Tafazolli (2016), power control techniques, whether centralised or distributed, are incapable of successfully dealing with network interference inflicted on devices participating in D2D. When it

Figure 3. Closed loop power control
Source: RF Wireless World, 2018

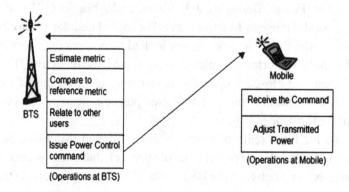

comes to the interference from CUEs to devices engaged in D2D communication, the researchers found that power control strategies don't help much. According to Xing and Hakola (2010), power control techniques must be used in conjunction with other interference mitigation schemes to be effective. Mode selection, link adaptability, and resource scheduling can all be employed with power control methods.

In D2D-enabled cellular networks where D2D would have been implemented to boost cellular network throughput, most power control systems that have appeared in literature are designed to cope with cross-tier as well as co-tier interference. According to Mahdi (2016), the power control approaches are intended to help D2D improve overall network throughput. The fundamental goal of these proposed D2D schemes is to shift mobile data traffic to nearby devices via resource sharing rather than the utilization of the BS. This is done to make better use of the frequency spectrum, resulting in higher overall throughput in each cell. According to Lopez (2016) and Lin, Andrews, Ghosh, and Ratasuk (2014), D2D has attracted the interest of many researchers, but there has been relatively minimal research on power control techniques for possible immobile D2D-enabled devices such as ProSe-enabled sensors that are fixed in one location. This is what sparked the idea for this study.

THE PROPOSED SCHEME

For this research work, it is proposed that a ProSe-enabled sensor should connect initially to a BS using the OLPC. According to 3GPP (2014), the maximum transmit power that it can use is:

$$P_{Tx} = \min\left(P_{CMAX}, P_{target} + PL\right)[dBm].$$ (1)

Where:

P_{CMAX} = The maximum allowable transmit power for the ProSe-enabled sensor.
P_{target} = target reception power.
PL = the pathloss between the BS and the ProSe-enabled sensor and is calculated in the ProSe-enabled sensor.

This transmit power is then corrected by the sensor after getting feedback from the BS to equation (2):

$$P\_Tx = \min \begin{cases} P_{CMAX(i)}, \\ P_0\left(j\right) + 10\log_{10}\left(2^{\mu}.M_{RB}\left(i\right)\right) + \alpha'\left(j\right).PL + \Delta_{TF} + f\left(i\right) \end{cases} \tag{2}$$

Where:

P_{CMAX} = Configured UE maximum power.

P_0 = The power to be contained in one RB. It is a cell specific parameter that is determined using higher layer parameters and measured in dBm/RB.

M = The number of allocated RBs per user.

μ = A 5G New Radio parameter which changes the sub-carrier spacing (SCS).

α' = This is a pathloss compensation factor which is a cell specific parameter in the range [0, 1].

Δ_{TF} = A correction function from the BS to the UE for fine tuning the in UE's transmission power.

PL = Downlink path-loss estimate in dB calculated by the UE using a reference signal.

$f(i)$ = Power control adjustment function.

Combining the power control adjustment and correction functions simplifies equation (2) to:

$$P_{Tx} = \min \begin{cases} P_{CMAX(i)}, \\ P_0\left(j\right) + 10\log_{10}\left(2^{\mu}.M_{RB}\left(i\right)\right) + \alpha\left(j\right).PL + \Delta_j \end{cases} \tag{3}$$

Taking a subcarrier spacing of 15kHz and using 1 RB, ($2^{\mu} = 1\left(for\,\mu = 0\right)$, M =1) simplifies equation (3) to:

$$P_{Tx} = \min \begin{cases} P_{CMAX(i)}, \\ P_0\left(j\right) + \alpha\left(j\right).PL + \Delta_j \end{cases} \tag{4}$$

Where Δ_j is the Transmit Power Command (TPC) generated by BS.

The TPC value is a correctional value that comes as feedback from the BS. For this research work, it is proposed that the TPC value should be obtained by making use of the received SINR and an SINR target. If P_{CMAX} is less than the calculated power, $P_0\left(j\right)+\alpha\left(j\right).PL+\Delta_j$, the sensor transmits to the BS using the calculated power, if it is more, then the sensor transmits at P_{CMAX}.

P_{CMAX} is 23dBm 2, which is Power Class 3(PC3) according to 3GPP (2018). High-performance user equipment (HPUE) is being proposed for 5G, which are devices capable of transmitting in the uplink with a power higher than 23dBm (MediaTek, 2018). This boost in uplink transmit power is very welcome, since it will allow ProSe sensors to cover a larger area, lowering the number of sensors required.

Once the transmit power that a sensor must use when communicating with a BS has been determined, the sensor and a nearby CUE must determine the transmit power that they must use once a BS has authorized D2D communication. The maximum transmit power of devices communicating directly with one another should be strictly managed, not only to avoid interference but also to ensure proper D2D communication. The objective of the proposed power control is to ensure that a network-prescribed SINR target for CUEs, $SINR_{ue}^{target}$, and a network prescribed SINR target for UE-Sensor (DUEs), $SINR_{due}^{target}$ are met. For a sensor that has to share an uplink resource with a CUE, the SINR over RB r at the BS is given by:

$$SINR_{ue}^r = \frac{G_{ueBS}.p_{ue}}{Ga + p_{D2D}G_{D_t.BS}} \tag{5}$$

And for the downlink, the SINR over resource block (RB) r at the CUE is given by:

$$SINR_{ue}^r = \frac{G_{ueD2D}.p_{D2D}}{Ga + p_{D2D}G_{D_t.BS}} \tag{6}$$

To meet pre-defined QoS requirements, this SINR over RB r should be equal to/or higher than a target SINR, $SINR_{ue}^{target}$ that is required at the BS for successful communication.

$$\frac{G_{ueBS}.p_{ue}}{Ga + p_{D2D}G_{D_t.BS}} \geq SINR_{ue}^{target} \tag{7}$$

If the interference that a sensor adds on RB r over subframe s is given by:

$$I_s^r = p_{D2D}G_{D_t.BS} \tag{8}$$

Then:

$$G_{D_tD_r}p_{D2D} \leq \frac{G_{cueBS} \cdot p_{ue}}{SINR_{ue}^{target}} - Ga \tag{9}$$

$$p_{D2D} \leq \frac{G_{cueBS} \cdot p_{ue}}{SINR_{ue}^{target}G_{D_tD_r}} - \frac{Ga}{G_{D_tD_r}} \tag{10}$$

Where:

Ga Represents the Gaussian noise power on a specific link.

p_{ue} Represents the transmit power of UE.

p_{D2D} Represents the transmit power of D2D (ProSe-enabled sensor to UE transmission).

$G_{D_tD_r}$ Represents channel gain between D2D transmitter (*Dt*) and D2D receiver (*Dr*).

G_{ueBS} Represents the channel gain between UE and the BS.

$G_{UE.D_r}$ Represents channel gain between UE and D2D receiver.

$G_{D_t.BS}$ Represents channel gain between D2D transmitter and the BS.

$SINR_{ue}^r$ SINR for any arbitrary UE in a cell over RB *r*.

$SINR_{D2D}^r$ SINR for any arbitrary D2D in a cell over RB *r*.

$P_{BS}G_{CU}$ Represents macro tier interference.

Algorithm 1 is used to control transmit powers of devices involved in a D2D communication.

Algorithm 1: Power Control Algorithm

1: Input: Matrix P // Paired UE- ProSe-enabled sensor devices
2: Output: A set of transmit powers, p = (p1, p2, . . ., pL);
3: for all pairs,
3.1: calculate transmit powers for the D2D transmitters using the equation:

$$p_{D2DTx} = \frac{G_{ueBS} \cdot p_{ue}}{SINR_{ue}^{target} G_{D_tD_r}} - \frac{Ga}{G_{D_tD_r}}$$

3.2: allocate powers accordingly
4: end for
5: $P = (p1, p2, \ldots, pL)$

SIMULATION AND DISCUSSION OF RESULTS

Radio propagation effects should always be adequately simulated, according to the research community, because these effects can significantly affect the system performance of any radio system. The propagation environment, noise, and interference are all factors that affect wireless communication networks. Signal attenuation is caused by large obstructions, as well as the distance between devices and multi-path fading. Pathloss, shadowing, and multipath fading are all factors that are considered in this research. The models can deal with events like the one depicted in Figure 4. Many people and items absorb radio energy in these settings, as well as building walls, highly polished floors and ceilings that deflect, absorb, and scatter the signals. These obstructions affect radio communications, including rapid fading and significant signal attenuation.

Deployment of Network Elements

For this study, 5 CUEs are dropped at random near a stationary ProSe-enabled sensor at a random distance from a BS. CUE installation and results collection are repeated several times until representative results are acquired. Figure 5 shows five CUEs that have been randomly placed in a 300m by 300m space. The suggested technique takes a snapshot of the CUEs' specific locations, as well as the BS and ProSe-enabled sensors' placements. The BS is believed to be capable of receiving channel quality and geolocation data. Because the 5 CUEs are so close to a ProSe-enabled sensor, they may be allowed to engage in D2D communication with ProSe-enabled sensor. Permission to engage in D2D communication is first sought from the BS, which grants permission if resources are available and the QoS conditions are met.

Figure 6 displays a 180m by 200m region with 5 CUEs dropped randomly 10 times around an immobile BS and a fixed ProSe-enabled sensor. As shown in Figure 6, the simulation is carried for multiple UE droppings in various locations. Convergence and statistically acceptable results can only be obtained with a large number of droppings.

Figure 4. Wernhill Park Shopping Mall, Central Windhoek, Namibia
Source: Mwashita, 2019

Figure 5. 5 CUEs, 1 ProSe-enabled sensor and a BS

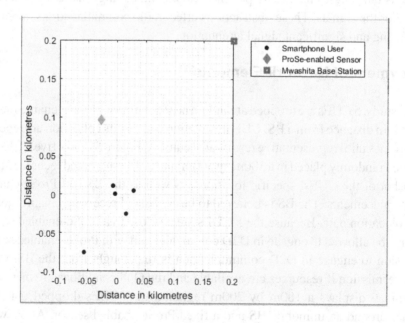

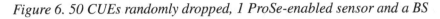

Figure 6. 50 CUEs randomly dropped, 1 ProSe-enabled sensor and a BS

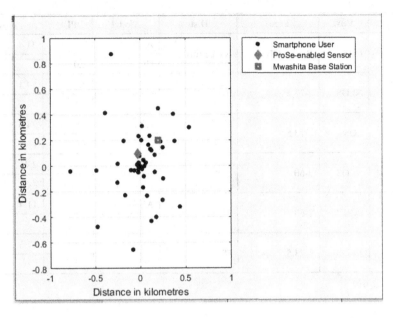

Path Loss Model

Sun *et al.* (2017) provided a route loss model for a 5G micro cellular scenario, which was used to evaluate the power control strategy described in this study. The alpha-beta-gamma (ABG) model was utilized first, followed by the close-in (CI) free space reference distance model. This is because the CI model is quite close to the existing 3GPP path loss model in terms of structure. A floating constant is substituted with a frequency dependent free space path loss in the CI model, which is based on a one-meter standard CI reference distance. This slight modification streamlines the analytical process, resulting in improved accuracy across a wide range of mmWave and microwave frequencies.

The traffic model proposed by Hossain (2013) is employed in this study because it produces a realistic traffic profile like that of a real cellular network. The traffic generation in practical mobile networks is inhomogeneous, with the arrival rate varying both non space and time. CUEs generate and forward high-quality level video streams to one another via the BS in this research effort, whereas ProSe-enabled sensors compete for resources for D2D communication with SUEs. The tests that were performed are listed in Table 3.

Table 2. Parameters in the ABG and CI path loss models in UMi and UMa scenarios

Scenario	Env.	Freq.	Dist.	Model	PLE	β(dB)
UMi SC	LOS	Range (GHz)	Range (m)	ABG	2.0	31.4
				CI	2.0	-
	NLOS	2-73.5	5-121	ABG	3.5	24.4
				CI	3.1	-
UMi OS	LOS	2-73.5	19-272	ABG	2.6	24.0
				CI	1.9	-
	NLOS	2-60	5-88	ABG	4.4	2.4
				CI	2.8	-
UMa	LOS	2-60	8-235	ABG	2.8	11.4
				CI	2.0	-
	NLOS	2-73.5	58-930	ABG	3.3	17.6
				CI	2.7	-

(Sun et al., 2016)

Table 3. Tests conducted

Test Item	Tests Conducted
1	Variation of ProSe-enabled sensor transmit power with distance
2	Reusable distance
3	Maximum allowed transmit power
4	ProSe-enabled sensor SINR variation with distance
5	Impact of D2D communication on CUE throughput
6	Impact of resource sharing on the SINR of one CUE

Simulation Parameters

The suitable settings for the simulation framework were chosen for the effective evaluation of the suggested technique for interference management. The simulation parameters proposed by the 3GPP in 2017 and 2018 for usage with 5G D2D systems were employed. Table 4 lists the parameters.

Table 4. Simulation parameters

Parameter	Value
System bandwidth	100 MHz
Carrier frequency	4 GHz
Maximum smartphone transmission power	23dBm
Maximum BS transmission power	24 dBm
Maximum ProSe-enabled sensor transmission power	26 dBm
Shadowing Standard Deviation	10 dB
Noise Spectral Density	-174 dBm /Hz
Total number of available RBs	50
D2D transmission power	Proposed power control scheme to meet a specified SINR
Monte Carlo simulation runs	1000
Bit rate requirement for prioritised CUEs	2Mbps
Bit rate requirement for non-prioritised CUEs	1.0Mbps

Variation of ProSe-enabled Sensor Transmit Power With Distance

Figure 7 depicts the transmission power that a ProSe-enabled sensor should consume at various distances from the BS for various levels of power regulation. It can be seen that when = 0, the sensor transmits at maximum power to the BS, and as the value grows, the transmit power required to communicate to the BS for the same distances reduces. This means that as increases, the amount of interference between cells using the same resources decreases. Figure 7 further shows that when the distance between the BS rises, sensors must transmit at increased power.

Reusable Distance

Figure 8 depicts the distance from the BS at which uplink (UL) resources currently in use by a certain CUE can be utilized by a ProSe-enabled sensor without compromising cellular network QoS. The reusable distance falls rapidly as the power control level, increases in both non-line of sight (NLOS) and line of sight (LOS) propagation conditions, as shown in Figure 8. It's also worth noting that in an NLOS propagation environment, the reusable distance reduces slightly faster than in a LOS propagation environment. It's also clear that ProSe-enabled sensors should never be allowed to

Figure 7. Variation of power with distance

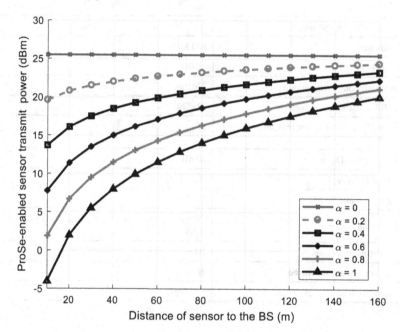

share UL resources with CUEs that are more than 160 meters away from the BS in an NLOS environment and 150 meters away in a LOS environment.

Maximum Allowed Transmit Power

Figures 9 and 10 show that at CUE distances closer to the BS, the maximum allowable power increases more quickly to reach the maximum power. For CUEs that are very close to the BS, reusing subchannels at distances greater than 115m from the BS for an NLOS environment should not be permitted.

ProSe-enabled Sensor SINR Variation With Distance

The SINR of a ProSe-enabled sensor varies with distance from a BS, as shown in Figure 11. SINR is high when a sensor is close to a BS and drops exponentially as the distance from the BS grows with a power control of 0.7. It can also be seen that at distances of up to 120m from the BS, LOS propagation loses its advantage over NLOS propagation in terms of SINR. This is a strange phenomenon that can be explained by the fact that many reflections and scatterings from ceilings and walls, which are abundant in densely built-up areas like shopping malls, as well as the

Figure 8. Reusable distance from BS vs power control levels

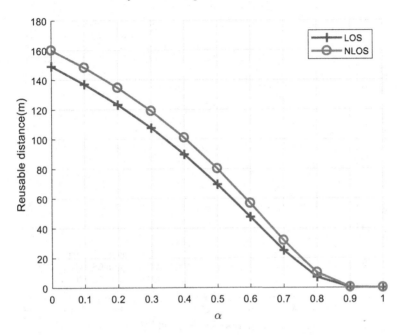

Figure 9. Maximum transmit power vs distance for LOS

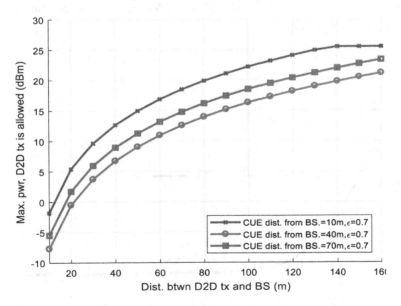

Figure 10. Maximum transmit power vs distance for NLOS

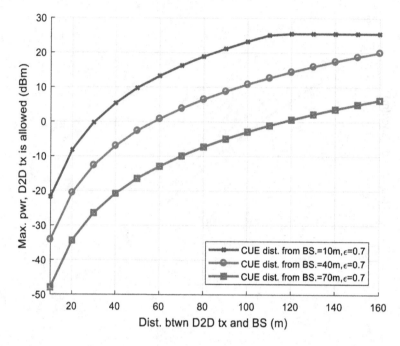

Figure 11. Variation of a CUE's throughput with distance from BS for $(\alpha = 2.0)$ *LOS and* $(\alpha = 3.1)$ *NLOS*

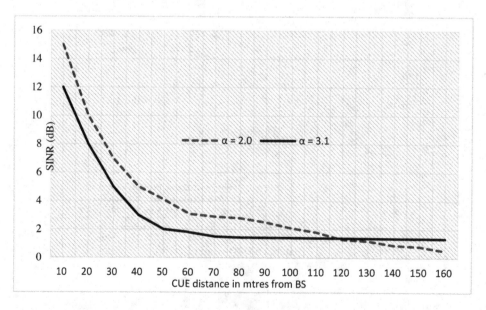

Figure 12. CUE 1's throughput before introducing D2D communication

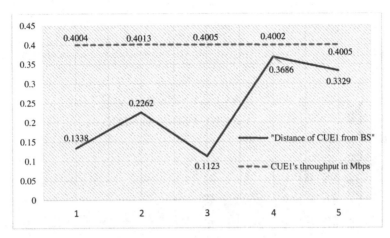

waveguiding effects of hallways, passages, and alleys, all contribute to an increased received power, (Rappaport, Xing, MacCartney, Molisch, Mellios and Zhang 2016),

Impact of D2D Communication on CUE Throughput

Figures 12 and 13 depict the difference in throughput of one CUE. Figure 12 depicts how the CUE's throughput changes as it traverses across the cell. The CUE's throughput is then tracked again once a ProSe-enabled sensor is allowed to

Figure 13. CUE 1's throughput after introducing D2D communication

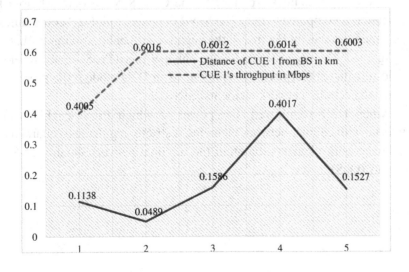

Figure 14. SINR of one CUE with and without resources being shared

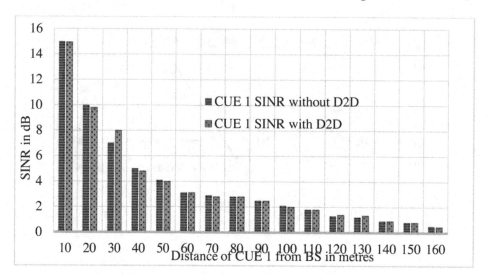

utilise the CUE's UL resources. The reuse of resources by the ProSe-enabled does not appear to have harmed the CUE's throughput, as seen in Figure 13. The minor increase in throughput at 4.8m from the BS can be attributed to the high SINR that signals enjoy over short distances from the BS, as seen in Figure 11.

Impact of Resource Sharing on the SINR of One CUE

The effect of allowing ProSe-enabled sensors to share UL resources on the CUE SINR is investigated by computing the variation of the CUE SINR with cell distance before UL resources are shared. The CUE SINR acquired before D2D communication is compared to the CUE SINR obtained after D2D communication is implemented. As seen in Figure 14, the negative influence is so minor that it can be safely ignored. The suggested technique only permits a ProSe-enabled sensor to communicate with cellular network components in D2D mode if the process does not degrade the cellular network elements' SINR.

Even when D2D is allowed, the power at which ProSe-enabled sensors communicate with cellular network elements is strictly regulated to guarantee that the cellular network elements maintain the minimal network Quality of Service requirements (QoS).

CONCLUSION

A power control method for ProSe-enabled sensors and smart phones connected to these sensors in a D2D fashion was developed in this research study. A fractional power approach with pathloss compensation is used in the proposed approach. The ability of ProSe-enabled sensors to link directly to smartphones and BSs is expected to degrade a cellular network's QoS. The proposed technique uses a target SINR to set the transmit powers of D2D devices, ensuring that the QoS of a cellular network is not jeopardized.

There hasn't been a state-of-the-art system that considers interference avoidance in 5G networks when smartphones interact with ProSe-enabled sensors directly. To test the method, a tiny system-level MATLAB emulator specifically tailored for 5G networks was created. The capabilities of the system-level simulator were as follows:

1. The capturing of the randomness of the distance between a ProSe-enabled sensor and the BS.
2. The capturing of the mobility of the CUEs.
3. Capturing the randomness of the location of ProSe-enabled sensors.

Six tests were done on the system level simulator to evaluate the effectiveness of the suggested method, and the results reveal that the proposed strategy is capable of appropriately managing interference levels that are known to cause problems in D2D-enabled networks. According to simulation data, allowing ProSe-enabled sensors to interact directly with neighbouring UEs can result in a 3.6 percent drop in overall user throughput. This value is modest, and both network consumers and network providers are likely to accept it. According to Ramasamy (2017:45), a 5 percent drop is acceptable to both users and network providers and is frequently reflected in service level agreements between network providers and their clients. When compared to standard power management strategies, the power control strategy described in this paper results in an acceptable drop in a cellular network's overall QoS.

The results show that the overall network QoS is not degraded by limiting the power at which ProSe-enabled devices transmit to the BSs and the power at which D2D devices talk with each other. This is significant because cellular network providers are less inclined to accept strategies that undermine network quality of service. The proposed technique has a flaw in that it does not account for the heterogeneity of next-generation mobile networks. As a result, it would be advantageous to expand the project in the future to include small cells, as well as to quantify and optimize the performance of the suggested technique in such networks.

REFERENCES

3rd Generation Partnership Project Organisational Partners. (2017). *Proximity-based services (ProSe), 3GPP TS 23.303 V15.0.0.* Valbonne: 3GPP Organisational Partners' Publications Offices.

3rd Generation Partnership Project Organisational Partners. (2018). *Physical layer procedures for control, (3GPP TS 38.213 version 15.2.0 Release 15),* Valbonne: 3GPP Organisational Partners' Publications Offices, 1-101.

3rd Generation Partnership Project Organizational Partners. (2014). *Technical Specification Group Services and System Aspects; Study on architecture enhancements to support Proximity-based Services (ProSe) (Release 12)," 3GPP TR 23.703 V12.0.0 (2014-02).* Valbonne: 3GPP Organizational Partners' Publications Offices.

Abdallah, A., Mansour, M. M., & Chehab, A. (2017). A Distance-Based Power Control Scheme for D2D Communications Using Stochastic Geometry. In *Proceedings of the 2017 IEEE 86th Vehicular Technology Conference (VTC-Fall).* Toronto: IEEE.

Bhushan, B., & Sahoo, G. (2018). Recent Advances in Attacks, Technical Challenges, Vulnerabilities and their Countermeasures in Wireless Sensor Networks. *Wireless Personal Communications, 98,* 2037–2077.

Chen, J., Liu, C., Li, H., Li, X., & Li, S. (2016). A categorized resource sharing mechanism for device-to-device communications in cellular networks. *Mobile Information Systems,* 1-10.

Cho, S., Choi, J. W., & You, C. (2013). Adaptive multi-node multiple input and multiple output (MIMO) transmission for mobile wireless multimedia sensor networks. *Sensors (Basel),* 3382–3401.

Ericsson. (2017). *5G Readiness Survey 2017: An assessment of operators' progress on the road to 5G.* Stockholm: Ericsson.

Feng, D., Lu, L., Yuan-Wu, Y., Li, G. Y., Feng, Y., & Li, S. (2013). Device-to-Device Communications Underlaying Cellular Networks. *IEEE Transactions on Communications, 61*(8), 3541–3551.

Fodor, G., Penda, D. D., Belleschi, M., Johansson, M., & Abrardo, A. (2013). A comparative study of power control approaches for device-to-device communications. In *Proceedings of 2013 IEEE International Conference on Communications (ICC).* Budapest: IEEE.

Hossain, M. F. (2013). *Traffic-driven energy efficient operational mechanisms in cellular access networks* (Doctoral dissertation). The University of Sydney, Sydney.

Institute of Electrical and Electronics Engineers. (2006). *Multi-hop relay system evaluation methodology (channel model and performance metric), IEEE C802.16j-06/013r3*. IEEE.

Lin, X., Andrews, J., Ghosh, A., & Ratasuk, R. (2014). An overview of 3GPP device-to-device proximity services. *IEEE Communications Magazine, 52*(4), 40–48.

Lopez, S. M. (2016). *An overview of D2D in 3GPP LTE standard.* Retrieved from http://d2d-4-5g.gforge.inria.fr/Workshop-June2016/slides/Overview_LTE_D2D.pdf

Mach, P., Becvar, Z., & Vanek, T. (2015). In-band device-to-device communication in OFDMA cellular Networks: A Survey and Challenges. *IEEE Communications Surveys and Tutorials, 17*(4), 1885–1922.

Mahdi, A. H. (2016). *The integration of device-to-device communication in future cellular systems* (Doctoral dissertation). Technischen Universität, Berlin.

MediaTek. (2018). *5G NR Uplink enhancements better cell coverage & user experience.* Author.

Mwashita, W., & Odhiambo, M. O. (2018). Interference management techniques for device-to-device communications. In P. K. Gupta, T. I. Ören, & M. Singh (Eds.), Predictive Intelligence Using Big Data and the Internet of Things (pp. 219-245). IGI Global.

Rappaport, T. S., Xing, Y., MacCartney, G. R., Molisch, A. F., Mellios, E., & Zhang, J. (2017). Overview of Millimetre Wave Communications for Fifth Generation (5G) Wireless Networks-With a Focus on Propagation Models. *IEEE Transactions on Antennas and Propagation, 65*(12), 6213–6230.

Rathee, D., Ahuja, K.& Nayyar, A. (2019). Sustainable future IoT services with touch-enabled handheld devices. *Security and Privacy of Electronic Healthcare Records: Concepts, paradigms and solutions,* 131-152.

Safdar, G. A., Ur-Rehman, M., Muhammad, M., Imran, M. A., & Tafazolli, R. (2016). Interference Mitigation in D2D Communication Underlaying LTE-A Network. *IEEE Access: Practical Innovations, Open Solutions, 4,* 7967–7987.

Song, L., Niyato, D., Han, Z., & Hossain, E. (2015). *Wireless Device-to-Device Communications and Networks.* Cambridge University Press.

Sun, S., Theodore, S. R., Sundeep, R., Timothy, A. T., Amitava, G., Istvan, Z., ... Jan, J. (2016). Propagation Path Loss Models for 5G Urban Micro- and Macro-Cellular Scenarios. In *Proceedings of the IEEE 83rd Vehicular Technology Conference (VTC Spring)*. IEEE.

Wireless World, R. F. (2018). *Difference between open loop power control vs closed loop power control*. Retrieved from https://www.rfwireless-world.com/Terminology/Open-Loop-Power-Control-vs-Closed-Loop-Power-Control.html

Xing, H., & Hakola, S. (2010). The investigation of power control schemes for a device-to-device communication integrated into OFDMA cellular system. In *Proceedings of 21st Annual IEEE International Symposium on Personal, Indoor and Mobile Radio Communications*. IEEE.

Chapter 5

IoT Ecosystem:
Challenges Due to SEP Litigation

Keerti Pendyal

https://orcid.org/0000-0001-6728-4157
O. P. Jindal Global University, India

ABSTRACT

In this chapter, the author looks at the challenges to the IoT system due to standard essential patents (SEPs) by looking at guidelines issued by regulators across the world to enable policymakers and judiciaries to deal with critical issues raised in cases involving SEPs. SEPs present a unique challenge as they require balancing the principles of intellectual property law and competition policy. The author analyses four critical challenges raised in disputes involving SEPs by looking at policy guidelines and arrives at the best practices drawn from these guidelines so that they may be used as guideposts for policymakers and regulators to resolve the increasing number of disputes involving SEPs. Finally, the author identifies some key challenges and systemic issues that are yet to be addressed – issues at the centre of some of the most significant disputes involving SEPs today.

INTRODUCTION

Patents have been accorded special protection under the law to allow the people who create it or generate it to gain financially through the use of the property provided the innovators/creators of the intellectual property place the innovations in the public domain. The law is formulated to strike a balance between the rights and responsibilities of innovators and the greater public good generated from the dissemination of the knowledge (created by the innovators). As a result, while at

DOI: 10.4018/978-1-7998-9312-7.ch005

the same time protecting the innovators from other imitators seeking to gain off of the inventions, the law also requires innovators to make the details of the innovation public. This is done to encourage knowledge development through new inventions based on the earlier innovations, whose inventors must disclose the details publicly in exchange for patent rights.

As described above, patent rights bestow a specific set of rights upon the innovators. These rights are negative. This means that the person/entity who owns these rights can prevent others from commercially exploiting or otherwise taking advantage of the assets covered by these rights without the explicit permission of the rights holder. In the case of patent rights, they can be the right to prevent commercialisation, copy, distribute copies, etc. These negative rights give the patent rights holder a monopoly over the commercialisation (or otherwise gaining monetarily) for a specific period. While the intent of this might seem to be at cross-purposes with laws that seek to curb monopolistic tendencies, both intellectual property laws and competition/antitrust laws share common goals – to promote innovation, enable efficient allocation of resources by the economy and improve consumer welfare. Regulatory agencies and the judiciary also recognise these common goals. In Atari Games Corp. v. Nintendo of Am., Inc., 897 F.2d 1572, 1576 (Fed. Cir. 1990), the court pointed out that

[T]he aims and objectives of patent and antitrust laws may seem, at first glance, wholly at odds. However, the two bodies of law are actually complementary, as both are aimed at encouraging innovation, industry and competition (Atari Games Corp. v. Nintendo of Am., Inc., 897 F.2d 1572, 1576 (Fed. Cir. 1990) c.f. U.S. Department of Justice, Federal Trade Commission, 2017, p. 2).

Although both intellectual property laws and competition laws have similar goals, intellectual property can, in some cases, give the I.P. rights owner market power, which the rights holders can abuse. This is especially true in the case of blocking patents. High-technology industries are greatly susceptible to the issue of blocking patents. Blocking patents are often a crucial part of new technological development or underpin an entirely new avenue for research. Given their nature, the holders of blocking patents wield an enormous amount of market power – they can easily stifle innovation (especially from competitors). Because of this, any new invention or research which requires licensing from the holder of blocking patents can come to a stop or become entirely unprofitable. Blocking patents can also run afoul of competition authorities, especially if the patent holder and its competitors are working on similar research. These outcomes are often displayed in sectors with a rapid rate of technological advancements and see quick obsolescence, like in the case of information and communication technology or high-tech industries. The power that the owner of blocking patents holds over other competitors or researchers is not something that was observed recently. The ability of the blocking patent holders to stifle innovation has been studied in depth since the 19th century, where it was first looked at in great detail by Cournot (Cournot 1838, c.f. Shapiro 2001). Cournot describes the issue of blocking patents itself as a special case of the complements problem that he studied.

As applied to the case of blocking patents, consider two firms, A and B, both making the same product P. The manufacture of P requires the use of patents X and Y with firm A owning X and firm B owing Y. Now, Cournot shows that the manufacturer of P would benefit the most if both the patents (X and Y) were held by a single firm. However, since one firm owns one patent each, the next best alternative for both firms is to cross-license their patent to the other firm to manufacture the product and benefit from the same. Solving the complements problem requires some degree of cooperation among the competing firms. Although it could be beneficial to the end-user, the collaboration among the competing firms seeking to solve the complements problem may face challenges from the competition authorities. The cross-licensing of blocking patents is one solution to the challenges posed by blocking patents. Some of the other solutions to get around the problem posed by Cournot and faced with blocking patents are package licensing, creation of patent pools and standard-setting. This chapter focuses on the issues arising from standard-setting as they apply to the IoT industry.

BACKGROUND

The Internet of Things refers to the ecosystem formed by the interconnected network formed through devices connected to a communication network – generally the internet. While the name strictly requires the interlinked devices to be connected to the internet and able to send/receive instructions and data through the internet, the name 'Internet of Things' has come to represent a connected network of devices that can 'talk' to each other, even if this network is not connected to the internet. The technology that allows hitherto 'dumb' devices to be connected to each other and the internet enables us to embed intelligence into these networks to enable automation and decision-making without human involvement. The evolution of technology and the advent of cheaper manufacturing (Oracle India 2020) have allowed for the proliferation of IoT devices in many industries worldwide. We are seeing a rapid explosion of connected devices in the consumer space with the adoption of devices like connected locks, thermostats, cameras, doorbells, home appliances (refrigerators, microwave ovens, vacuum cleaners, etc.), automobiles (Graham, 2019); Mueller, 2021; Mueller, 2021; Mueller, 2021), and personal health devices (insulin pumps, pedometers, fitness bands, etc.). IoT devices in the consumer space have to be compliant with several standards depending on the platform they use to connect to the internet. Some of these are standards that have been around since before the widespread adoption of the Internet of Things – standards like WiFi and Bluetooth. At the same time, other standards were developed for exclusive use in the IoT ecosystem, like the Zigbee and Z-wave standards (Gallego & Drexl, 2019). In addition to the widespread adoption of the Internet of Things in the consumer space, we are also seeing the adoption of IoT enabled equipment in the industrial area, with machinery in several industries being equipped with sensors so that companies/manufacturers can track the performance of the equipment, ensure optimal performance and suggest preventive action/maintenance to reduce downtime.

We also see rapid adoption of IoT in the B2B space with the advent of connected buildings (allowing for the remote monitoring of installations), connected supply chains, etc. The predominant standard for industrial IoT and B2B IoT is the Modbus protocol initially developed in 1979. In each of the communications protocols discussed above, through which the IoT devices talk to one another and are connected to the internet (WiFi, Bluetooth, Modbus, Thread, WirelessHART or the others), there are several patents involved which are quite often owned by different companies/parties (Gallego & Drexl, 2019). As a result, no single entity would be able to manufacture an IoT product without obtaining a license from all the patent holders/owners. This could give rise to the complements problem discussed in earlier paragraphs, thus preventing companies from making or deploying these solutions.

As explained above, one of the approaches that can be taken to resolve the complements problem is through standard-setting. In this scenario, the participating firms are part of an organisation that sets the standards for all the firms that are its members[1]. As part of its charter/mandate, the standard-setting organisation requires all the firms that hold patents essential to the standard to license these patents. The patent holder will license the patents on a FRAND basis – Fair, Reasonable and Non-Discriminatory (Arseven, 2021; Gallego & Drexl, 2019; Geradin & Katsifis, 2021; McDonagh & Bonadio, 2019; Podszun, 2019; Ungerer, 2021). The standard-setting approach also enjoys an advantage over the cross-licensing, package licensing and patent pool approaches since the rules of the standard-setting organisation require all essential patents to be licensed. The benefits of standard-setting are successful launching of a bandwagon or network, greater realization of network effects, protecting buyers from stranding, and enabling competition within an open standard (Shapiro 2001, 138).

The standard-setting process sometimes gives rise to a scenario where the holder of one of the patents which are part of the standard ends up wielding significantly more power than the patent holder (at the time of filing the patent) or the standard-setting organisation (at the time of standard formation) envisaged. This is due to the nature of the patent. Each standard has a set of core technical specifications and additional optional specifications that a product/service that claims to comply with the standard must fulfil. As a result, the patents which cover the technical innovations that are part of the core technical specifications need to be licensed by any manufacturer looking to develop a standard-compliant product (assuming the product does not include any additional optional specifications which are part of the standard). As a result, if a company wants to build products/services compatible with these standards, they have no workaround but to license these patents. These patents which a company has to license to comply with the standard are called Standard Essential Patents. Defined formally, a Standard Essential Patent (SEP) is a patent essential to implementing a standard agreed upon by the industry body or the Standard Setting Organisation. In other words, it is a patent without infringing (assuming the manufacturer doesn't license the patent) on which the standard cannot be implemented. SEPs provide a unique challenge to competition authorities and policymakers because the holder of an SEP is left with market power that is not representative of the value of the patent on its own. While the holder of an SEP legally has a right to stop others from infringing on his patent, sometimes such actions might run afoul of competitive policy or end up affecting innovation in an industry/across industries.

Although standardisation is a workaround to resolve some tricky issues involving blocking patents and the complements problem, it sometimes clashes with other economic policy goals. As we mentioned above, the process of standard-setting can attract attention from the competition authorities. It is also seen that the market power of companies holding Standard Essential Patents (SEPs) increases by a significant amount after the patents have been declared Standard Essential. While the patent holder has the right to stop third parties from infringing on its patent and seek damages from such behaviour (if infringement has occurred), courts in various jurisdictions have differed from each other as to the nature of the remedies available to the patent holder to seek a resolution of such dispute. There is also considerable divergence in opinion on the meaning of the terms 'Fair, Reasonable and Non-Discriminatory' when it comes to the commitments by the holder of the SEPs to license these patents on a FRAND basis.

The challenges involving SEPs has been brought into stark focus with recent cases involving Apple and Qualcomm (Porter, 2019) as well as FTC and Qualcomm (Mickle, Kendall and Fitch, 2019), along with industry actions involving Standard Setting Organisations (Huawei barred by both the S.D. Association and the WiFi Alliance both of which are Standard Setting Organizations. Huawei's membership in both the organisations was restored later in the week) (Keane, 2019; Gartenberg, 2019a; Gonzalez, 2019). The settlement of their lawsuit by Apple and Qualcomm directly resulted in one of Qualcomm's competitors (Intel) exiting the industry (Gartenberg,

2019b). This example shows the possible effect cases involving standards can have on competition and consumer welfare.

Like any other property dispute, patent disputes are essentially disputes between two private entities. However, although effectively property disputes, they gain some peculiarity by the nature of the property in question. Given the intangible nature of patents and the (generally) time-limited nature of ownership of these patents, the disputes are in a class unto themselves and differentiate themselves from property disputes in general. Additionally, cases involving Standard Essential Patents are further complicated by the vexatious nature of the intersection between property law, intellectual property law and competition law.

While any case has the potential to become a public policy issue (especially in a common law country like India), it is much more probable for cases involving alleged infringement of Standard Essential Patents to set the precedence for future judgements. It is only in recent years that the Indian judiciary has had to deal with cases that involve the alleged infringement of SEPs (there have only been a handful of cases that involve Standard Essential Patents filed in India). As a result, the judgements pronounced in the first few cases are crucial as the judicial logic and interpretation are seen as the benchmark/framework for future cases. Additionally, owing to the highly specialised nature of the subject matter involved and required (legal expertise, technical expertise and economic understanding to say the least), the judgements pronounced can have a far-reaching impact, possibly far beyond the case being determined.

The unique challenges that Standard Essential Patents pose to promote the efficient allocation of resources are recognised by countries worldwide. Several countries have come out with policy documents and guidelines to serve as markers for companies to navigate the issues of licensing and antitrust when it comes to Standard Essential Patents. These guidelines also aid policymakers and regulators in determining if the actions of companies are violating any of the I.P. laws or competition laws. This chapter will look at some of the challenges/issues arising out of the intersection of I.P. rights and competition policies. The major issues we identify in this chapter at this intersection were also identified in cases involving SEPs in India. We shall also be looking at how five different jurisdictions – the United States, Japan, the European Union, Canada and South Korea – approach these challenges by studying policy documents/guidelines issued by regulators in these territories. We will be analysing these guidelines and policy statements to identify potential best practices, which can then be used as a template for the competition authorities and patents offices in other countries worldwide. Finally, we shall try to see if the guidelines issued in these five countries could help resolve some of the questions raised in these cases.

POLICY GUIDELINES ON SEP ISSUES

Policymakers and regulators worldwide have looked at some of the antitrust issues that may arise from the enforcement of rights granted by patent laws. As a result, they have issued guidelines and policies to advise companies on avoiding some of these potential issues. These guidelines also help the judiciary understand the stance of the policymakers and regulators when adjudicating disputes involving these issues. Some of the large economies which have issued these guidelines are the United States, the European Union, Japan, South Korea, Canada, and China. As discussed in the previous section, we shall be looking at the guidelines issued by the U.S., E.U., Japan, South Korea, and Canada in this chapter – guidelines addressing potential antitrust issues in the licensing of intellectual property in general as well as specifically in the case of Standard Essential Patents.

United States

The United States was the first major economy to come out with policy guidelines concerning antitrust issues in licensing of intellectual property (U.S. Department of Justice, Federal Trade Commission, 2017) as well as guidelines covering issues concerning the licensing of Standard Essential Patents (U.S. Department of Justice, U.S. Patent & Trademark Office, 2013). The guidelines concerning antitrust issues in licensing of intellectual property were first published in 1995 and were updated in 2017. Although the United States was the first major economy to come out with guidelines concerning issues concerning the licensing of Standard Essential Patents, the U.S. Department of Justice later walked back from the position taken in these guidelines leading to the U.S. Department of Justice, the U.S. Patent & Trademark Office, and the National Institute of Standard and Technology issuing a modified policy statement on remedies for Standard Essential Patents subject to voluntary F/RAND commitments (U.S. Patent & Trademark Office; U.S. Department of Justice; National Institute of Standards and Technology; 2019).

Japan

Like in the case of the United States, Japanese regulators also issued two documents with guidelines on the use of intellectual property rights and licensing negotiations involving standard essential patents. These guidelines were issued by the Japan Fair Trade Commission and the Japan Patent Office, respectively (The Japan Fair Trade Commission, 2016; Japan Patent Office, 2018).

European Union

The European Commission released its Communication from the Commission to the European Parliament, The Council and the European Economic and Social Committee, setting out the E.U. approach to the Standard Essential Patents on 29 November 2017. Like the guidelines issued by the Japan Patent Office, this document seeks to set out key principles that foster a balanced, smooth and predictable framework for SEPs (European Commission, 2017, p. 2).

Canada

The Competition Bureau of Canada published the Intellectual Property Enforcement Guidelines to articulate how the Bureau approaches the interface between competition policy and I.P. rights (Competition Bureau Canada, 2016, p. 2).

These guidelines help the users to understand the circumstances in which the bureau would initiate action to investigate conduct involving I.P. rights and the circumstances where it would recommend/initiate legal action.

South Korea

The Fair Trade Commission of the Republic of Korea came out with Review Guidelines on Unfair Exercise of Intellectual Property Rights in 2016 (Korea Fair Trade Commission, 2016). These guidelines became effective on 23 March 2016, and their purpose was to provide a framework that can be used by the Commission to regulate action by intellectual property rights holders (both domestic and foreign enterprises as long as the actions/contracts of these enterprises affected the Korean market). The guidelines seek to help the Commission develop criteria that would enable it to determine if enterprises' exercise of intellectual property rights constitutes an abuse of dominance of market power or cartelisation by a group of companies.

IDENTIFICATION OF ISSUES AND ANALYSIS

As we saw in the previous sections, regulatory agencies worldwide have developed policies that deal with the antitrust issues in the licensing of intellectual property. Some of these agencies have also issued further guidelines covering the specific scenario of antitrust issues in the licensing of Standard Essential Patents. This section will look at these agencies' common arguments on different contentious problems that courts have had to decide on in cases involving Standard Essential Patents. These issues were highlighted in the lawsuits between SEP owners and allegedly infringing parties in lawsuits brought before the Indian judiciary. Briefly, these issues are as follows:

1. Which is the appropriate jurisdiction to decide on matters involving Standard Essential Patents? Is it the courts as described in most intellectual property laws, or do the antitrust regulators (Competition Commission of India in the case of India) have jurisdiction to adjudicate these disputes?
2. What is the appropriate way of determining royalty in cases involving Standard Essential Patents – Entire Market Value (EMV) basis or the Smallest Saleable Patent Practicing Unit (SSPPU) basis?
3. When is a licensor/licensee an unwilling licensor/licensee?
4. Is seeking an injunction against an allegedly infringing company by the owner of an SEP an anti-competitive practice?

Jurisdiction

As mentioned earlier, several lawsuits were filed in India by owners of SEPs starting in 2009. The first of these cases, filed by Philips, was decided in July 2018. Apart from Philips, lawsuits were filed by Ericsson, Dolby, and Vringo against various manufacturers of telecommunication and electronic equipment. Along with the lawsuits filed by Ericsson against different mobile phone manufacturers in India, it also filed a lawsuit against the Competition Commission of India. Ericsson's suit against the Competition Commission of India (CCI) was filed after the CCI issued orders directing the Director-General of CCI to begin an investigation against Ericsson for its alleged abuse of dominant position. These orders were issued based on complaints filed with the Competition Commission of India by three different mobile handset manufacturers (Micromax, Intex and iBall) (In Re: Micromax Informatics Limited And Telefonaktiebolaget LM Ericsson (Publ), 2013; In Re: Intex Technologies (India) Limited And Telefonaktiebolaget LM Ericsson (Publ), 2014; In Re: M/s Best I.T. World (India) Private Limited (iBall) And M/s Telefonaktiebolaget L M Ericsson (Publ), 2015) each of whom is individually locked in a patent infringement suit with Ericsson. The point of contention in Ericsson's lawsuit against the CCI was the jurisdiction of CCI to issue an order launching the investigation into Ericsson's behaviour. Ericsson contended that CCI lacked jurisdiction to pass such an order since the subject matter dealt with patents and that the Patents Act had listed the appropriate forums where cases dealing with patents could be raised.

In their complaints to the CCI, Micromax, Intel, and iBall had alleged that Ericsson was abusing its dominant position as one of the world's largest telecommunication companies and as one of the largest holder of SEPs in the mobile phone and wireless industries (In Re: M/s Best I.T. World (India) Private Limited (iBall) And M/s Telefonaktiebolaget L M Ericsson (Publ), 2015, p. 3).

More specifically, the complaints included:

1. Linking of royalty to the final price of the finished product (as against the functionality of the patent in the finished product).
2. Not sharing information about which patents were infringed without an NDA being signed.
3. The requirement to sign an NDA that would prevent the complainant from discussing the components that infringed on the patents with its suppliers. The complainants alleged that an NDA would also not have allowed them to check if the licensing offer made by Ericsson was FRAND in nature.
4. Imposing a jurisdiction clause in the NDA and the final licensing agreement insisting on disputes being adjudicated in a country different from where both the parties were doing business (India).

After going through the information provided by the three companies in their respective complaints, the CCI arrived at a preliminary finding that there was prima facie evidence that Ericsson had abused its dominant power. It directed the Director-General of the CCI to undertake an in-depth investigation into each of these complaints[2]. Ericsson filed petitions with the Delhi High Court arguing that the Competition Commission of India lacked jurisdiction to commence any proceeding in relation to a claim of royalty by a proprietor of a patent (Telefonaktiebolaget LM Ericsson (Publ) vs Competition Commission of India and Another, 2016, p. 4; Telefonaktiebolaget LM Ericsson (Publ) vs Competition Commission of India and Another, 2016, p. 4).

Ericsson argued that any dispute concerning royalty would fall entirely within the purview of the Patents Act, 1970 and cannot be examined under the Competition Act, 2002. As a result, such disputes would be beyond the jurisdiction of the Competition Commission of India.

In delivering a judgement in the initial cases filed by Ericsson, the judge looked at the legislative reasoning behind the enactment of both the Patents Act and the Competition Act. While he agrees that the Patents Act does provide potential licensees with remedies in the case of complaints like excessive licence fee, unreasonable and anti-competitive licensing terms, and breach of FRAND obligations (Telefonaktiebolaget LM Ericsson (Publ) vs Competition Commission of India and Another, 2016, p. 104; Telefonaktiebolaget LM Ericsson (Publ) versus Competition Commission of India and Another, 2016, p. 104), he also finds that the remedies granted by the two statutes are not mutually exclusive, i.e. remedies obtained under one statute do not prevent a party from seeking remedies under the other statute as well. The judge holds that since, in his opinion, there is no conflict between the Patents Act and the Competition Act, the jurisdiction of the CCI to entertain complaints for abuse of dominance in respect of Patent rights cannot be ousted (Telefonaktiebolaget LM Ericsson (Publ) vs Competition Commission of India and Another, 2016, p. 129-130; Telefonaktiebolaget LM Ericsson (Publ) vs Competition Commission of India and Another, 2016, p. 129-130).

The judge also held that although Ericsson's suits filed against Micromax and Intex were proceeding, the CCI could go ahead with its investigation. This order by Justice Vibhu Bakhru in these suits is critical. It enshrines the right of the Competition Commission of India to investigate anti-competitive behaviour even when it pertains to disputes involving patent licensing and royalties. Ericsson has since appealed the decision in these cases, and the appeals are currently being heard.

The plea by Ericsson that the Competition Commission of India lacked jurisdiction to entertain complaints against itself and the counter-claim by the CCI that it could do so have been addressed in the policy guidelines that are the focus of this chapter. The guidelines being analysed all state that while disputes involving patents primarily come under the jurisdiction of the courts who are granted the authority by the I.P. laws, in cases where there seems to be an abuse of antitrust law in the exercise of rights granted by the I.P. laws, these disputes can be addressed by antitrust regulators.

In Antitrust Guidelines for the Licensing of Intellectual Property, while recognising that the possession of intellectual property might create market power, both the agencies argue that by itself, this does not raise antitrust issues. At the same time, they also state that because intellectual property is governed under separate statutes, it does not exempt them from scrutiny under the antitrust laws (U.S. Department of Justice, Federal Trade Commission, 2017, p. 3) (nor does it mean that they are more likely to be scrutinised). The agencies argue that although some characteristics of intellectual property might be different from other forms of property, the fundamental principles of antitrust analysis are robust enough to deal with these differences with minor modifications to the framework of analysis. They also lay down directions to help one understand if a particular scenario will be treated as per se anti-competitive or not (called the rule of reason treatment in the guidelines).

Similarly, in the Guidelines for the Use of Intellectual Property under the Antimonopoly Act, the Japan Fair Trade Commission cautions users that sometimes these intellectual property rights may also end up curbing competition (The Japan Fair Trade Commission 2016). It then goes on to define the scope of the guidelines:

1. If the conduct arising from the exercise of intellectual property rights were to inhibit any other party from using technology
2. If the holder of the intellectual property were to license the technology with a very restricted scope

3.   If the holder of the intellectual property imposes restrictions/conditions on activities of the entities which take a license to use the technology

In the same way, the Intellectual Property Enforcement Guidelines published by Competition Bureau Canada clarify that the circumstance under which the bureau would investigate actions would fall into two broad categories – actions involving just the exercise of I.P. rights, and activities involving more than mere exercise of the I.P. right (Competition Bureau Canada, 2016, p. 6).

The guidelines also clarify that the bureau would not presume that the actions would violate the Competition Act in either of the cases mentioned above. Finally, the guidelines state that the bureau uses the same analytical framework to analyse conduct involving intellectual property as it does in the case of conduct that does not involve intellectual property. The bureau (through these guidelines) also lays down the enforcement principles which guide it in determining if action needs to be taken in situations involving intellectual property. While doing this, the bureau uses its framework to determine if there are any anti-competitive effects of the action being investigated, along with looking at probable efficiency considerations and other business justifications.

The Review Guidelines on Unfair Exercise of Intellectual Property Rights published by the Fair Trade Commission of the Republic of Korea (Korea Fair Trade Commission, 2016) seek to help the Commission develop criteria that would enable it to determine if the exercise of intellectual property rights by enterprises constitutes an abuse of dominance of market power or cartelisation by a group of companies. The Commission clearly states that the guidelines shall only be applicable when a company with market dominance exercises its intellectual property rights (In particular, refusal to trade, discriminations, and imposition of considerably excessive amount of royalty all by a company alone is, in principle, subject to this guideline only when the company has overwhelming market dominance) (Korea Fair Trade Commission, 2016, p. 5).

However, the Commission cautions that the mere possession of market dominance does not by itself constitute a violation of the Monopoly Regulations and Fair Trade Act. The guidelines also state that when evaluating the actions of enterprises, the anti-competitive nature of these actions needs to be compared to the increase in efficiency as a result (if any). Only if the anti-competitive effect outweighs the efficiency increases does the action violate provisions of the fair trade act.

From the above, we can see that regulators from different jurisdictions uniformly believe that in cases where the exercise of the rights granted by intellectual property law can be detrimental to competition, the antitrust authorities have the jurisdiction to investigate these actions. These guidelines also caution that the users of the policy documents should be cautious in this approach as the exercise of rights granted by I.P. law in itself may not be anti-competitive.

Appropriate Method of Determining Royalties

In the previous two sections, we observed how regulators worldwide affirmed the common purpose behind intellectual property law and antitrust law. We also saw how the regulators believed that in certain specific scenarios (defined in the guidelines), the antitrust authorities might have jurisdiction to look into disputes involving intellectual property. Each of the developed economies studied has developed policy documents/guidelines addressing the potential antitrust issues arising from licensing intellectual property. However, out of the five jurisdictions studied, only three – Japan, European Union, and the United States – have come up with specific guidelines dealing with issues arising out of licensing Standard Essential Patents (the United States, which was one of these three economies has since released an updated policy document which has in effect walked back the suggestions of the earlier document released in the year 2013. Additionally, the guidelines issued by the United States in 2013 were focused on the right to seek exclusion of the patent owners).

In this section, we shall look at the second major issue courts face in adjudicating disputes involving SEPs, viz., what is the appropriate method of determining royalty? Should the royalty be based on the market value of the finished product (the Entire Market Value or EMV approach), or should it be based on the smallest component of the product that is independently sold and which contains the patented invention (the Smallest Saleable Patent Practising Unit or SSPPU approach)? There are arguments in favour of both approaches (Arseven, 2021), and courts have sided with one argument over the other in the past (sometimes within the same country).

While the guidelines issued by both Japan and the European Union talk about appropriate royalty, they do so from a more generic perspective without choosing one method of determining royalty over the other. The guidelines issued by the European Union argue that the royalties should be reasonable (although they do not define what would be a reasonable amount), avoid royalty stacking, and explain as to why the royalty is a reasonable rate (providing examples if possible by referring to a third-party determination of the royalty). The European Commission calls on the parties to consider the following factors while valuing patents:

1. Licensing terms have to bear a clear relationship to the economic value of the patented technology (European Commission, 2017, p. 6)
2. Determining a FRAND value should require taking into account the present value added of the patented technology (European Commission, 2017, p. 7)

3. FRAND valuation should ensure continued incentives for SEP holders to contribute their best available technology to standards (European Commission, 2017, p. 7)
4. To avoid royalty stacking, in defining a FRAND value, an individual SEP cannot be considered in isolation. Parties need to take into account a reasonable aggregate rate for the standard, assessing the overall added value of the technology (European Commission, 2017, p. 7)

Similarly, the Guide to Licensing Negotiations involving Standard Essential Patents describes the different royalty calculation methods – reasonable royalties, non-discriminatory royalties and other ways of calculating royalties – as well as some additional factors that need to be considered in the calculation of royalties (number of licensees that have agreed to the royalty rate, the scope of the license, essentiality/validity/infringement of patent, the value of the individual patents, negotiating history and volume discounts) (Japan Patent Office, 2018). However, the guidelines are silent on which approach to determining royalty is a more suitable approach leaving it instead to the negotiating parties (or the courts) to decide.

Unwilling Licensor/Licensee

Similar to the issue of the appropriate method of determining royalties, only the guidelines issued by the European Union and Japan look at the issue of unwilling licensor/licensee. Although both the policy documents do not provide any absolute criteria to determine if a company is an unwilling licensor/licensee, they provide some steps that a company can follow not to be classified by courts (if the negotiations break down) as an unwilling participant in the negotiations.

In the Guide to Licensing Negotiations involving Standard Essential Patents, when elaborating on the licensing negotiation methods, the regulator (Japan Patent Office) puts forward two crucial factors that play a role in negotiations. One is good faith, and the other is efficiency. The guidelines state that when a license is sought on FRAND terms, the phrase 'Fair, Reasonable and Non-Discriminatory' applies not just to the rate of royalty but also to the parties' behaviour during the negotiation. The guidelines also lay down the steps in an ideal good faith negotiation (although it cautions that not every negotiation needs to have all the steps in the process or the same order):

1. Licensing Negotiation Offer from Rights Holder (Japan Patent Office, 2018, p. 7)
2. Expression from Implementer of Willingness to Obtain a License (Japan Patent Office, 2018, p. 7)

3. Specific Offer from Rights Holder on FRAND Terms (Japan Patent Office, 2018, p. 7)
4. Specific Counteroffer from Implementer on FRAND Terms (Japan Patent Office, 2018, p. 7)
5. Rejection by Rights Holder of Counteroffer/Settlement of Dispute in Court or through ADR (Japan Patent Office, 2018, p. 7)

Similar to the steps highlighted by the Japan Patent Office for a negotiation to be seen as a good-faith negotiation, the guidelines also lay down some key factors to be considered for the negotiations to be conducted efficiently. These are:

Notification of a Timeframe. (2018). (pp. 22–23). Japan Patent Office.
Parties to Negotiation in Supply Chain. (2018). (pp. 22–23). Japan Patent Office.
Protecting Confidential Information. (2018). (pp. 22–23). Japan Patent Office.
Choice of Patents subject to Negotiation. (2018). (pp. 22–23). Japan Patent Office.
Geographic Scope of License Agreement. (2018). (pp. 22–23). Japan Patent Office.
Patent Pool Licensing. (2018). (pp. 22–23). Japan Patent Office.
Greater Transparency of SEPs. (2018). (pp. 22–23). Japan Patent Office.

Similar to the guidelines issued by the Japan Patent Office, the European Commission also lays down some general principles for parties engaging in taking FRAND licenses for SEPs. It calls on all parties to the negotiation to engage in good faith. While determining the royalty rates for the SEPs involved, the Commission argues that what constitutes fair, reasonable and non-discriminatory would depend on the sector involved (of both the SEP holder and the potential licensee).

Seeking Injunctions: Anti-competitive in Nature?

The fourth major issue that regulators and courts face is whether seeking injunctions by SEP owners is anti-competitive. The right to exclude is one of the core rights granted to intellectual property owners. This includes the right to prevent commercialisation, copy, print, distribute copies, etc. Intellectual property rights essentially give the rights holder a monopoly over the product for a specific period by providing negative rights. Intellectual property owners are given these rights to protect the innovators from other imitators seeking to gain off the inventions.

However, in their guidelines, the European Commission and Japan Patent Office caution the owners of SEPs against indiscriminately seeking injunctions. Citing the judgement in the Huawei vs ZTE case (Case C-170/13 Huawei Technologies, E.U.:C:2015:477 c.f. European Commission, 2017), the Commission is of the opinion that SEP holders should *not seek injunctions against users willing to enter into a license on FRAND terms (European Commission, 2017, p. 9).*

In scenarios where the courts determine that granting an injunction is appropriate, the Commission requires the injunctive relief granted to be

effective, proportionate and dissuasive (European Commission, 2017, p. 10).

Similarly, in the guidelines issued by the Japan Patent Office, the regulator recognises that companies providing a product or service using SEPs are faced with the threat of injunctions resulting in a 'hold-up' situation. At the same time, it also observed that legal precedents across the world seem to be converging toward permitting injunctions concerning FRAND encumbered SEPs (i.e., SEPs for which a FRAND declaration has been made) only in limited situations (Japan Patent Office 2018, 1). The regulator also points out that the decision on whether an injunction is justified or not very often depends on the behaviour of the two parties in the negotiating process, thus linking this issue with the previous issue discussed regarding an unwilling licensor/licensee.

In the context of injunctions, it is important to look at the Policy Statement on Remedies for Standard Essential Patents issued jointly by the U.S. Department of Justice and the U.S. Patent & Trademark Office in 2013. In this document, the agencies sought to address a particular issue in cases involving standard essential patents – whether injunctions or other similar exclusion orders can be issued in cases involving standard essential patents, especially when these patents are covered by a voluntary F/RAND commitment given by the patent holder. Despite acknowledging that there is a possibility of misuse of exclusion orders being sought by the patent holder, the Department of Justice and the U.S. Patent & Trademark Office conclude that injunctions and exclusion orders are an appropriate remedy available to innovators patent holders to exercise their rights. At the same time, they also recommend caution in granting these orders.

Although it was one of the first countries to come out with a policy statement covering the potential issues involving the licensing of standard essential patents, the agencies involved with the Policy Statement on Remedies for Standard Essential Patents have since issued an updated policy statement in this regard(U.S. Patent & Trademark Office; U.S. Department of Justice; National Institute of Standards and Technology; 2019). The revised policy guidelines essentially reverse the position of the agencies and state that the earlier version of the policy statement has led to misinterpretation suggesting that a unique set of legal rules should be applied in disputes concerning patents subject to a F/RAND commitment that are essential to standards (as distinct from patents that are not essential), and that injunctions and other exclusionary remedies should not be available in actions for infringement of standards-essential patents (U.S. Patent & Trademark Office; U.S. Department of Justice; National Institute of Standards and Technology; 2019, 4).

In the updated statement issued jointly by the U.S. Department of Justice, US Patents & Trademark Office, and the National Institute of Standards and Technology, the agencies clarify that . . . a patent owner's F/RAND commitment is a relevant factor in determining appropriate remedies, but need not act as a bar to any particular remedy (U.S. Patent & Trademark Office; U.S. Department of Justice; National Institute of Standards and Technology; 2019, 4). They further state

All remedies available under national law, including injunctive relief and adequate damages, should be available for infringement of standards-essential patents subject to a F/RAND commitment, if the facts of a given case warrant them. Consistent with the prevailing law and depending on the facts and forum, the remedies that may apply in a given patent case include injunctive relief, reasonable royalties, lost profits, enhanced damages for willful infringement, and exclusion orders issued by the U.S. International Trade Commission. These remedies are equally available in patent litigation involving standards-essential patents (U.S. Patent & Trademark Office; U.S. Department of Justice; National Institute of Standards and Technology; 2019, 5).

BEST PRACTICES AND LESSONS

In the last decade, numerous cases have been filed in India involving Standard Essential Patents (SEP holders have filed 21 lawsuits against companies alleging infringement. This is not counting the cross complaints filed by the alleged infringing companies against the SEP holders in courts and the Competition Commission of India). While these cases have raised several issues concerning whether or not the actions of the implementing companies amount to infringement of the SEP holders' rights, there are several other issues (that were listed in an earlier section) that the author feels are more systemic and which can be addressed partly by looking at the points raised in the guidelines and policy documents discussed so far.

Common Focus of I.P. Laws and Antitrust Laws

The most important lesson that we can draw from the policy documents analysed in this chapter is the common focus of the I.P. laws and antitrust laws. Regulators from across the different jurisdictions studied – both intellectual property authorities and antitrust regulators – are uniform in their belief that the purpose of both I.P. laws and antitrust laws is to increase the efficiency of allocating economic resources.

While they do acknowledge that sometimes, in specific instances, these laws may be at cross-purposes, the fundamental goal behind enacting these statutes remains the promotion of the efficient functioning of the economy.

In Antitrust Guidelines for the Licensing of Intellectual Property (these guidelines which were issued in 2017 replaced the earlier *Antitrust Guidelines for the Licensing of Intellectual Property* guidelines issued on 6 April 1995), the United States Department of Justice and the Federal Trade Commission lay out some basic principles before looking at the unique issues that may involve intellectual property. Most importantly, they affirm that both intellectual property laws and the antitrust laws share the common purpose of promoting innovation and enhancing consumer welfare (U.S. Department of Justice, Federal Trade Commission, 2017, p. 2).

The purpose of intellectual property laws is to provide innovators with several rights, including the right to exploit the innovation commercially and prevent others from performing specific actions, including copying, profiting from the innovation, etc. These rights are granted to innovators both for engaging in innovation and making public the invention, which leads to the increase of our cumulative knowledge. Without the innovators being given these rights, imitators would likely be able to rapidly exploit the innovations by copying them, thus reducing the incentive for creators and innovators to engage in innovation. In the long run, this is detrimental to consumer welfare. Similarly, antitrust laws promote innovation and consumer welfare by prohibiting certain actions that may harm competition with respect to either existing or new ways of serving consumers (U.S. Department of Justice, Federal Trade Commission, 2017, p. 2).

Similar to the antitrust guidelines for the licensing of intellectual property in the United States, these guidelines re-affirm the importance of intellectual property for entrepreneurs to engage in research and development and the procompetitive effects of intellectual property.

Like in the case of the Japanese and American guidelines/policy documents, the Intellectual Property Enforcement Guidelines issued by the Competition Bureau of Canada also begin by reiterating the importance of innovation in the economy today. The guidelines also highlight the importance of intellectual property laws and competition laws to promote an efficient economy. The purpose of these guidelines was to articulate how the Bureau approaches the interface between competition policy and I.P. rights (Competition Bureau Canada, 2016, p. 2).

These guidelines help the users to understand the circumstances in which the bureau would initiate action to investigate conduct involving I.P. rights and the circumstances where it would recommend/initiate legal action.

Although the policy documents issued by the European Union and the Republic of Korea do not explicitly mention the common focus of intellectual property laws and antitrust laws, this aspect and reasoning of the respective regulatory agencies becomes visible through the way the regulators proposed investigating possible antitrust issues. This shows that both intellectual property authorities and antitrust regulators viewed the I.P. laws and antitrust laws as having a common focus of increasing the economic efficiency of the resources developed and deployed in the market.

Jurisdiction

Among the identified issues, the most important question is the jurisdiction of the Competition Commission of India (the statutory regulator tasked with implementing the competition laws in India) in adjudicating complaints concerning the exercise of intellectual property rights. In three of the cases filed by Ericsson against Indian mobile handset manufacturers, the defendants approached the Competition Commission of India (CCI) alleging anti-competitive behaviour by Ericsson (Competition Commission of India, 2013; Competition Commission of India, 2014; Competition Commission of India, 2015). Ericsson, in turn, pleaded in the Delhi High Court that the CCI lacked jurisdiction to look into these complaints (High Court of Delhi at New Delhi, 2016a; High Court of Delhi at New Delhi, 2016b). However, this argument of Ericsson was struck down by the Delhi High Court in its order dated 30 March 2016. The judge declared that the CCI could investigate actions involving the exercise of intellectual property rights to see if they were anti-competitive. The guidelines issued by the USA, Japan, Canada, and the Republic of Korea, which were analysed in this chapter, all state that in situations where the exercise of rights granted by I.P. laws can have an anti-competitive effect, the antitrust regulators of the countries had jurisdiction to investigate actions of enterprises. However, the regulators advise caution on behalf of the users of these documents – judiciary and companies – by stating that although the antitrust regulators may have jurisdiction in disputes involving intellectual property, they would do so in an extremely limited set of circumstances. While these documents are not binding, they help clarify the position of the antitrust agencies and the logic/framework that the agencies could use in analysing enterprises' actions. Such a document would also help the enterprises as it would remove uncertainty concerning the policy and the implementation of the same in the country.

Inter-agency Cooperation and Clarity in Policies

Aside from the issue of jurisdiction and the mutual focus of both I.P. laws & antitrust laws, the biggest takeaway is the cooperation between regulatory agencies tasked with applying different statutes. In addition to issuing policy documents, the antitrust agencies in the countries studied came out with these guidelines (and others) in cooperation with the respective intellectual property agencies. This inter-agency cooperation would further help reduce policy uncertainty and assist the judiciary in understanding the logic behind the provisions of these guidelines. Such collaboration would also prevent dissonance between the views of the regulatory agencies. This would be a critical step that needs to be taken to break down the silos in which different agencies work and would also help in the development of policies that are not contradictory.

Additionally, enterprises involved in SEP license negotiations would benefit significantly if the guidelines laid down the best practices that enterprises can follow during the negotiation process. These best practices could include the factors to be considered by the enterprises to engage in good faith negotiations, factors to be considered when efficiency in negotiation is of more importance, etc. These best practices would also help the enterprises understand how long they need to attempt negotiations before seeking judicial remedies and arbitration to resolve the disputes.

KEY CHALLENGES REMAIN

While there are several steps that policymakers can take to simplify the implementation of policy and learn best practices from their peers around the world, there are several key challenges that remain. We shall look at two of the biggest challenges here.

One of the most significant issues facing the judiciary in disputes involving patents is the appropriate method of determining royalty. Courts have taken differing views in different cases – sometimes choosing the EMV approach as the more appropriate method of determining royalty and in other cases choosing the SSPPU approach. There have been instances where different courts in the same country have differed on the proper method of determining royalties. In such a scenario, the issue of the appropriate method needs to be looked at more closely. The matter of calculating royalties was only considered by the Guide to Licensing negotiations involving Standard Essential Patents (Japan Patent Office, 2018). However, even these guidelines only briefly consider the pros and cons of each method of calculating royalties.

The authors of a report commissioned by the European Parliament to study the recommendations of the European Commission's communication on Standard Essential Patents conclude that the findings and recommendations made by the European Commission in its communication are to a great extent reasonable and appropriate (McDonagh & Bonadio, 2019, p. 30).

They also argue that the approach laid down by the Court of Justice of the European Union (CJEU) in its Huawei v ZTE judgement and followed by other judiciaries in subsequent years balance the interests of all stakeholders. They also argue that guidelines laid down by the European Commission in the communication about the procedure to be followed in determining the FRAND rates of royalty on a sound basis (McDonagh & Bonadio, 2019). However, the authors of the report argue that the European Union institutions can take several steps to make the licensing of SEPs better suited to the IoT industry, like increasing the viability of patent pools, collective licensing schemes, and increasing the viability of 'open source' approached in standardization processes (McDonagh & Bonadio, 2019, p. 31)

Gallego and Drexl argue that due to the nature of the IoT industry and the connectivity options required by IoT devices, the IoT market is considerably different from the mobile telephony market (Gallego & Drexl, 2019). As a result of this, they argue that the principles and frameworks used by authorities (specifically the authorities in the European Union on whom their paper is focussed) will need to be modified – especially the principles and frameworks laid down by the Court of Justice of the European Union (CJEU) in its judgement in the Huawei v. ZTE case (Huawei Technologies Co. Ltd v ZTE Corp. and ZTE Deutschland GmbH 2015 c.f. Gallego & Drexl, 2019). They predict that regulators and the judiciary at the national level in different countries will face challenges going ahead in the IoT industry owing to the complexity of the sector, the environment, and the number of players involved being considerably larger than in the mobile telephony space (Gallego & Drexl, 2019).

The methodology of the royalty determination is one of the issues raised by the counter complaints filed by the Indian mobile handset manufacturers with the CCI (Competition Commission of India, 2013; Competition Commission of India, 2014; Competition Commission of India, 2015). The royalty pricing methodology was also one of the issues at the heart of the lawsuit filed by the U.S. Federal Trade Commission against Qualcomm. This question – basis for different methods of calculating royalties, pros and cons of each, and applicability of different methods – needs further research to be resolved, and proactive action by the regulatory agencies will help us gain clarity in this regard sooner.

In addition to the questions raised in the lawsuits in India (regarding the jurisdiction, royalties, unwilling licensor/licensee, and seeking injunctions), there is one additional issue that has not been raised in these lawsuits that needs to be studied as well. Is the owner of an SEP who did not commit (to the Standard Setting Organisation) to licensing their patents on a Fair, Reasonable and Non-Discriminatory (FRAND) basis required to license these patents on a FRAND basis? Are they engaging in anti-competitive practices by not licensing the patents on a FRAND basis? This question is one of the issues at the heart of the FTC lawsuit against Qualcomm in the United States, which has since been appealed (Mickle, Kendall and Fitch 2019) and overturned. The resolution of this question will need all the stakeholders coming together to resolve it.

REFERENCES

Arseven, M. (2021). *Standard Essential Patents and Their Role in Enabling the Internet of Things.* https://www.lexology.com/library/detail.aspx?g=b614f8c7-0d02-4dd3-869e-9c1a83e30d7e

Competition Bureau Canada. (2019). *Intellectual Property Enforcement Guidelines.* Author.

Competition Commission of India. (2013). *Micromax Informatics Limited And Telefonaktiebolaget LM Ericsson.* Author.

Competition Commission of India. (2014). In Re: Intex Technologies (India) Limited And Telefonaktiebolaget LM Ericsson (Publ). Author.

Competition Commission of India. (2015). In Re: M/s Best IT World (India) Private Limited (iBall) And M/s Telefonaktiebolaget L M Ericsson (Publ). Author.

European Commission, (2017). *Communication from the Commission to the European Parliament, the Council and the European Economic and Social Committee - Setting out the E.U. approach to Standard Essential Patents.* Author.

Gallego, B. C., & Drexl, J. (2019). *IoT Connectivity Standards: How Adaptive is the Current SEP Regulatory Framework?* (Vol. 50). IIC - International Review of Intellectual Property and Competition Law.

Gartenberg, C. (2019). *Huawei can't officially use microSD cards in its phones going forward.* Academic Press.

Gartenberg, C. (2019). *Intel says Apple and Qualcomm's surprise settlement pushed it to exit mobile 5G.* Academic Press.

Geradin, D. & Katsifis, D. (2021). *End-product- vs Component-level Licensing of Standard Essential Patents in the Internet of Things Context.* Academic Press.

Gonzalez, O. (2019). *Huawei gets double bad news from S.D. Association and WiFi Alliance.* Academic Press.

Graham, S. (2019). *Nokia, Daimler, Continental Ramp Up Global Patent Chess Match.* Academic Press.

High Court of Delhi at New Delhi. (2015). *Telefonaktiebolaget LM Ericsson (Publ) vs Competition Commission of India and Another.* Author.

High Court of Delhi at New Delhi. (2016). *Telefonaktiebolaget LM Ericsson (Publ) vs Competition Commission of India and Another.* Author.

Japan Patent Office. (2018). *Guide to Licensing negotiations involving Standard Essential Patents.* Author.

Keane, S. (2019). *Huawei membership restored by S.D. Association.* WiFi Alliance.

Korea Fair Trade Commission. (2016). *Review Guidelines on Unfair Exercise of Intellectual Property Rights.* Author.

McDonagh, L., & Bonadio, E. (2019). *Standard Essential Patents and the Internet of Things: In-Depth Analysis.* Academic Press.

Mickle, T., Kendall, B., & Fitch, A. (2019). *Qualcomm's Practices Violate Antitrust Law.* Judge Rules.

Mueller, F. (2021a). *Sisvel becomes third Avanci licensor to sue Ford Motor Company over cellular standard-essential patents.* Academic Press.

Mueller, F. (2021b). *L2 Mobile Technologies claims Qualcomm chips in Ford, Lincoln cars infringe 3G standard-essential patents originally obtained by ASUSTeK.* Academic Press.

Mueller, F. (2021c). *Japanese patent licensing firm I.P. Bridge is suing Ford Motor Company in Munich over former Panasonic SEP.* Academic Press.

Oracle India. (2020). *What is IoT?* Author.

Podszun, R. (2019). Standard Essential Patents and Antitrust Law in the Age of Standardisation and the Internet of Things: Shifting Paradigms. *IIC - International Review of Intellectual Property and Competition Law, 50*, 720-745.

Porter, J. (2019). *Apple will try to tear apart Qualcomm's biggest business in court this week.* Academic Press.

Shapiro, C. (2001). Navigating the Patent Thicket: Cross Licenses, Patent Pools, and Standard Setting. In A. B. Jaffe, J. Lerner, & S. Stern (Eds.), *Innovation Policy and the Economy* (pp. 119–150). MIT Press.

The Japan Fair Trade Commission. (2016). *Guidelines for the Use of Intellectual Property under the Antimonopoly Act.* Author.

Ungerer, O. (2021). FRAND in IoT ecosystems. *Intellectual Property Magazine,* (July/August), 60–61.

U.S. Department of Justice, U.S. Patent & Trademark Office. (2013). *Policy Statement on Remedies for Standards-Essential Patents Subject to Voluntary F/RAND Commitments.* Author.

U.S. Department of Justice, Federal Trade Commission. (2017). *Antitrust Guidelines for the Licensing of Intellectual Property.* Author.

U.S. Patent & Trademark Office, U.S. Department of Justice, & National Institute of Standards and Technology. (2019). *Policy Statement on Remedies for Standards-Essential Patents Subject to Voluntary F/RAND Commitments.* Author.

ENDNOTES

[1] Standards can also be formed through the functioning of the marketplace without the involvement of a Standard Setting Organization (SSO) sometimes also called Standard Development Organization (SDO). In this scenario, a product becomes so widely adopted that its specifications become accepted as the accepted standard for the development of competing products or complements. It is also possible that there is competition in the marketplace between competing products and one of these products emerges as the standard which is adopted by the entire market. The example of Betamax vs. VHS is one such scenario where the Betamax format and the VHS format competed in the market with VHS emerging as the established standard.

[2] The complaint filed by iBall was withdrawn by the company after Ericsson and iBall settled their lawsuit and entered into a Global Patent License Agreement on 20th October 2015. As a result, both Ericsson and iBall petitioned the Delhi High Court to direct CCI to drop the investigation CCI initiated acting on the complaint of iBall (Telefonaktiebolaget LM Ericsson (Publ) vs Competition Commission of India and Anr 2015).

Chapter 6
Secure Privacy–Oriented Location–Aware Network

Ambika N.

(iD) https://orcid.org/0000-0003-4452-5514

St. Francis College, India

ABSTRACT

IoT is an amalgamation of diverse devices. The system aims to overcome the infrastructure of the devices. The instruments communicate with each other to accomplish a task. The previous contribution aims in preserving privacy among the communicating devices. It supports formation of cluster, where the device chosen as a cluster head mediates the communication. The user with the devices will be able to post their queries to the untrusted server by camouflaging themselves. The untrusted server responds to the queries which are communicated to the users through the cluster head. Security is vital to the network. The attacks if detected at an early stage can conserve large amount of energy. The current proposal works to enhance reliability to the network by 4.96%, 1.31% enhancement in detecting the attacks, and conserves energy by 6.13% compared to the previous contribution.

INTRODUCTION

Internet-of-things (Khan & Salah, 2018) (Ambika N., 2020) is the assembly of devices of divergent calibre. They (Ambika N., 2021) talk to one another using a common platform. They are used in enormous applications – home surveillance (Pokhrel, Vu, & Cricenti, 2019), healthcare (Purri, Choudhury, Kashyap, & Kumar, 2017), industry (Chen, Xu, Liu, Hu, & Wang, 2014) etc. to name a few. IoT (Nagaraj, 2021) is also used in area-based surveillances.

DOI: 10.4018/978-1-7998-9312-7.ch006

LOCATION-BASED SERVICES (LBS)

Area-based administrations are known as location-based services (Junglas & Watson, 2008) (Küpper, 2005). They provide assistance accounting for the geographic area. The system provides benefits to the manufacturer and purchaser. It gives clients the administrations guidelines. The administrations offer the likelihood to clients or machines. It finds different characters, devices, carriers, assets. The solicitation can begin from the customer and another substance like an application supplier or the network. The client needs to give authorization for the area demand. Some of its properties are –

- There is consistently a minimum of two substances engaged with an area-based help demand at any rate.
- One of the entities can be either static or moving.
- It is possible that they are by nature static, or they are just incidentally motionless.
- One of the elements is consistently the object of interest (regardless of whether human or non-human).
- The system records the doings.

Applications of LBS

- Savvy transport frameworks (Tomatis, Cataldi, Pau, Mulassano, & Dovis, 2008) are creating an innovative vision for data the combination among the scope of associations and administrations dynamic in transport arranging and tasks. These frameworks are alluded to as astute because their capacities permit them to perform higher-request activities, for example, situational examination and versatile thinking. The key innovations that empower the system vision are – geo-positioning, remote correspondences, adaptable figuring stages, and spatial data sets.
- Area-based games (Yu, 2008) characterize as PC games in which this present reality area of the player impacts how the competition creates. It partitions into ones got from open-air and table sports. The game portrays as joining outside exercises like chasing, covering up, or pursuing. It uses extra game components given by versatile innovation. It depends on collaboration and correspondence.
- The frameworks help outwardly impeded individuals with walker routes. It covers both arranging a strategy to a predetermined objective just as managing the client along the course. MoBic (Petrie, Johnson, Strothotte, Raab, Fritz, & Michel, 1996) is a comparative sort of help framework.

CONTRIBUTION OF THE PROPOSED TECHNOLOGY

The previous contribution (Alrahhal, Alrahhal, Jamous, & Jambi, 2020) aims to preserve privacy by camouflaging the cluster concept. The devices commute through the cluster head with the untrusted clients. The proposed arrangement relies upon the cooperative relationship abusing the common advantage among creatures. The projection of the shared merit is incredible trust between the individuals from a bunch and the pioneer. The group individuals will have the option to try not to associate with the LBS worker. The pioneer will misuse the genuine questions with authentic situations as fakers to pick up full security insurance at his/her side. The LBS worker is mostly answerable for security approach execution, while the client's mission sends his/her question. This gathering expects that the LBS worker should be dependable. The security strategy executes on the cell phones of the LBS clients. The LBS worker views as a malignant segment. The putting away geographic guide alongside annoyance-based assurance settles the security issue on the client-side. It is a different network length. Hilbert bend planning changes over the two measurements of the put-away guide into one measure. In the third gathering, the client either settles on her security assurance choice dependent on the assistance given by the LBS worker side or associates with the LBS worker for the situation where no inquiry answer is in stock. The LBS worker is viewed as an assailant moreover. LBS worker reactions are stored to get profits by noting approaching questions with the advancement of time. The clients of every cell will be gathered in one group. A pioneer chooses a bunch. If a question-answer is not available in the stock, the backer will send the inquiry to the pioneer. The developer sends it to the untrusted LBS worker. The pioneer gets the appropriate response after controlling the question on the LBS worker side. It afterward restores the got answer to the needed client. The developer abuses genuine questions based on undisputed positions as fakers on his side. At that point, the Leader will send his question with the authentic situation with no compelling reason to deliver fakers either at the area or inquiry level. LBS worker reactions are stored to acquire profits by noting approaching inquiries with the advancement of time. In situations where no answer is available in the store, the LBS client compels to interface with the untrusted LBS supplier.

The contributions of the paper-

- The reliability is enhanced by incorporating hashing concept. The devices suffix the hashed value using the identification, location details, and time of transmission. The hashed code cannot duplicate as the time and location vary with time.
- The attacks are detected at an early stage. The algorithm verifies the freshness w.r.t time.

- The previous work aims in preserving the privacy of the devices by camouflaging using cluster concept. To enhance the previous doing, the current proposal aims in securing the network, increasing reliability, and conserving the energy of the network.

The work divides into seven sections. The literature section details the contribution made by diverse authors. The proposal is explained in segment three. The proposal explains in the fourth division. The details of the simulation are jolted in the fifth section. The sixth section analyses the work. The work concludes in seventh section.

LITERATURE SURVEY

This segment narrates diverse contributions. This work (Alrahhal, Alrahhal, Jamous, & Jambi, 2020) aims to preserve privacy by camouflaging the cluster concept. The devices commute through the cluster head with the untrusted clients. The proposed arrangement relies upon the cooperative relationship abusing the common advantage among creatures. The projection of the shared merit is incredible trust between the individuals from a bunch and the pioneer. The group individuals will have the option to try not to associate with the LBS worker. The pioneer will misuse the genuine questions with authentic situations as fakers to pick up full security insurance at his/her side. The LBS worker is mostly answerable for security approach execution, while the client's mission sends his/her question. This gathering expects that the LBS worker should be dependable. The security strategy executes on the cell phones of the LBS clients. The LBS worker views as a malignant segment. The putting away geographic guide alongside annoyance-based assurance settles the security issue on the client-side. It is a different network length. Hilbert bend planning changes over the two measurements of the put-away guide into one measure. In the third gathering, the client either settles on her security assurance choice dependent on the assistance given by the LBS worker side or associates with the LBS worker for the situation where no inquiry answer is in stock. The LBS worker is viewed as an assailant moreover. LBS worker reactions are stored to get profits by noting approaching questions with the advancement of time. The clients of every cell will be gathered in one group. A pioneer chooses a bunch. If a question-answer is not available in the stock, the backer will send the inquiry to the pioneer. The developer sends it to the untrusted LBS worker. The pioneer gets the appropriate response after controlling the question on the LBS worker side. It afterward restores the got answer to the needed client. The developer abuses genuine questions based on undisputed positions as fakers on his side. At that point, the Leader will send his question with the authentic situation with no compelling reason to deliver fakers either at the area or inquiry level. LBS worker reactions are stored to acquire profits by noting approaching inquiries with the advancement of time. In situations where no answer is available in the store, the LBS client compels to interface with the untrusted LBS supplier.

The LBS framework (Sun, Liao, Li, Yu, & Chang, 2017) comprises three parts. It has a dispersed design without including a trusted anonymizer. The LP works as per applicable guidelines and arrangements in the LBS framework. With the security boundaries, the PIDS bootstraps the LBS framework and instates it. On the off chance that the client needs to be served by the LBS framework, he needs to enroll himself in the LBS framework. For sending solicitations to the LP, he will first and foremost join with different clients by utilizing the total convention. The client needs to settle on a choice if the clients have a similar area name. If they are touchy areas, the LLB calculation calls the pseudo-ID trading convention. If they are usual areas, the LLB calls the improved PLAM convention. Though the k clients have area marks, it straightforwardly calls the improved PLAM convention.

The LBS worker (Sun, et al., 2017) is dependable to get administration inquiries from clients, look for mentioned administration information in the data set, and answer with the list items back to the clients. The LBS worker can get the worldwide data dependent on inquiries. It can be the recorded question probabilities of clients identified. The framework comprises clients who outfit with cell phones. It is with work in GPS modules to get the client's area information. The ADLS calculation initially gets the secrecy degree as indicated by the client's area data. At that point, for the area in the client's area data, the calculation chooses other sham areas voraciously dependent on entropy and afterward gets the faker area set parameter. After getting the faker area sets, the computation sorts the probabilities of set boundaries, the client's spurious area collections in the climbing request. For each fake domain set to them, the ADLS calculation computes the change between the sets and the client's spurious area.

Protection estimation and measurement plan figures are dependent on the factual and data hypothetical models. The goal (Ukil, Bandyopadhyay, & Pal, 2014)is to get security measures from the credential guideline. It is the disambiguation of protection measures among security conservation ensures like k-namelessness. The measurable connections are registered. This improves security estimation and measurement precision. It performs the Kolmogorov-Smirnov test. It is nonparametric theory verification. It assesses the Cumulative Distribution Functions. The result of security measures contrasts with the REDD dataset. The adequacy of the proposed plan estimates the security hazard likelihood with standard disaggregation or Non-Intrusive Load Monitoring. Protection measurement empowers the clients to conclude whether to share their private sensor information with outsider applications.

In the first stage (Bahirat, He, Menon, & Knijnenburg, 2018), the work build up a layered settings interface. The clients settle on a less granular level and possibly move to a more granular choice when they want more itemized control. It diminishes the unpredictability of the options clients need to make. It happens without decreasing the measure of control access to them. The authors use a factual investigation of the Lee and Kobsa dataset. The viewpoint is the most elevated layer of our IoT protection setting interface. The perspectives consigned to bring down layers. They build up a brilliant default setting. It acquires the requirement for some clients to change their settings. Analysts have had the option to set up particular bunches or "profiles" given client conduct information. They perform AI examination on the Lee and Kobsa dataset to make a comparative arrangement of "keen profiles" for the IoT protection setting interface. Every member has fourteen situations depicting a circumstance where an IoT gadget would gather data about the member. It was a blend of five relevant limits, controlled at a few levels utilizing a blended fragmentary factorial plan that permitted us to test principal impacts and two-route collaborations between all boundaries.

In light of the hypothetical and procedural reflections, the creators (Chanson, Bogner, Bilgeri, Fleisch, & Wortmann, 2019) plan the exploration venture in three plan cycles, each made out of five stages. It trails the two last strides of assessment and correspondence. The planning cycle starts with a writing audit to distinguish the current issue and consider the plan. The assignment sets off with a report on the pervasiveness of odometer extortion. The inferred schema standards are in the target definition stage and recognize the design. It includes that are needed to address these plan standards. They started up the created plan concerning a particular use case and built up the first form of our model CertifiCar. The assessment adjusts the relic plan in the subsequent plan cycle and, given these changes, executed another rendition of ancient rarity. It incorporated the discoveries into the third plan cycle. It evaluates the outcome of the subsequent plan cycle. The certificate is in a field test with a hundred vehicles, and the ensuing assessment depended on the consequences of this field test and master interviews. They iteratively refined the plan necessities, standards, and highlights, improving the effectiveness of the outcome. They accumulate extra cuts of information for an itemized ex-post assessment of the determined plan necessities. It also includes standards and highlights of antiquity.

The confirmation can happen at four focuses on this model (Alpar,et.al, 2016) Client-controlled sensors are wearable and brilliant home gadgets that can speak with one another locally. For this situation, the confirmation finds on traits. When speaking, the sensors ought to conform to the administration. After assortment and preparing, information processors ought to impart limited data after confirmation of the accepting party. It gives another accreditation to an element on innovation. To qualify a component for the specific certification, the accreditation supplier needs to confirm the entity. It should be possible with regular distinguishing proof or by trait-based confirmation. The maker of an IoT gadget makes a keen understanding. It offers capacities for dealing with the ownership move and surveying the ownership of the device. The producer sends the agreement in blockchain and implants the location. The blockchain can be public, kept up by its locale, including its engineers, clients, specialist organizations, diggers, and others. At last, the maker moves the possession to the principal proprietor of the gadget.

The agreement (Islam & Kundu, 2018) comprises code and information. The capacities are transfer, and poll. The state factors are proprietors, inhabitant, occupant public key, and gadget shared solution. As the agreement is changeless whenever it is onto the blockchain, nobody can mess with the code. To move the tenure to an occupant, the proprietor sends an exchange to the ownership Contract. This exchange characterizes all the vital data identified with a tenure exchange, for example, new occupant data, occupancy period, and cost. It incorporates the inhabitant's public key. It processes the information encryption key. The smart contract refreshes its new inhabitant and the new occupant's shared solution. The IoT gadget surveys the location of the implanted brilliant agreement discontinuously. The encryption motor of the TPM changes the encryption key to the recently processed symmetric key. At that point, it encodes all the video information transfer or another payload with the encryption key. The occupant can likewise unscramble the video information with the shared key. Then again, the proprietor doesn't decode the observation information as the credential changes.

The researchers (Naeini, et al., 2017) directed an inside subject's review with 1,014 Amazon Mechanical Turk 1 laborers to comprehend people's security inclinations. They presented every member to fourteen distinct vignettes introducing an IoT information assortment situation. They changed eight factors that we conjectured could impact people's protection inclinations. Every vignette presented the elements in a similar request. In every condition, vignettes started with the information assortment and finished with the maintenance period. They built five factual models. It includes comfort level, permits for the information assortment, and intimation.

To add a gadget to the intelligent home, the excavator (Dorri, et al., 2017) creates a beginning exchange by offering a key to the device utilizing summed up Diffie-Hellman algorithm. The divided key among the excavator and the gadget is put away in the beginning exchange. Concerning characterizing the arrangement header, the property holder produces its approaches. It adds the approach header to the first square. The excavator utilizes the approach header in the most recent chain in the blockchain. It refreshes the arrangement the proprietor should refresh the most recent square's approach header. The devices may discuss straightforwardly with one another or with elements outer to the brilliant home. It is inside the home demands information from another inner gadget to offers some administrations. A shared key allotment by the excavator to the instrument straightforwardly speaks with one another. To dispense the key, the excavator checks the arrangement header or requests authorization from the proprietor and afterward circulates a divided solution among gadgets. After getting the credential, devices impart straightforwardly as long as it is substantial. The excavator denotes the disseminated key as invalid by sending a control message to gadgets. The advantages of this technique are twofold. The excavator has a rundown of gadgets that share information, and on the other, the interchanges between gadgets are made sure about with a shared key. The device should confirm the capacity to store data locally. The device needs to send a solicitation for the digger, and off chance, it put away consent. The excavator creates a shared key and sends the key for the gadget and the capacity. By accepting the key, the nearby stockpiling produces a beginning stage that contains the shared key. The instrument can store information straightforwardly in the neighborhood capacity. The gadgets store information on the distributed device. The requester needs a beginning stage that contains a square number and a hash utilized for mysterious validation purposes. In the wake of getting a solicitation, the capacity makes a beginning stage and sends it to the digger. When a gadget needs to store information on the distributed storage, it sends information and solicitation to the excavator. By getting the solicitation, the digger approves the device for putting away information on the distributed store. The excavator removes the last square number and hash from the nearby chain, and makes a store exchange, and sends it alongside the information to the capacity. After putting away learning, the distributed storage restores the new square number to the excavator. It is utilized for additional putting away exchanges. The excavator sends the current information of the mentioned gadget to the requester after accepting a screen exchange. If a requester is permitted to get information, then the digger sends knowledge occasionally until the requester sends a nearby solicitation to the excavator and cancels the exchange. The screen exchange empowers property holders to watch cameras or different gadgets in which send occasional information.

The library (Datta, Apthorpe, & Feamster, 2018) comprises systems administration natives that shape traffic to circulations free of client exercises. It performs traffic forming in the sending course by cushioning payloads, dividing payloads, and adding cover parcels. The cover traffic comprises irregular bytes. The send work adds 6-7 bytes of overhead showing fracture and load cushioning subtleties. It permits the getting capacity to dispose of cover traffic and recuperate messages in their unique structure. The library muddles client movement since payload sizes. The inter-packet spans draw from foreordained circulations free of client exercises.

PREVIOUS CONTRIBUTION

As the number of systems is increasing, how they perform may not be up-to the mark. Hence these systems use unregistered servers. Using this facility they will be able to accomplish the task. The systems require privacy as they do not want to reveal their identities to the untrusted server.

The previous contribution (Alrahhal, Alrahhal, Jamous, & Jambi, 2020) aims to preserve privacy by camouflaging the cluster concept. The devices commute through the cluster head with the untrusted clients. The proposed arrangement relies upon the cooperative relationship abusing the common advantage among creatures. The projection of the shared merit is incredible trust between the individuals from a bunch and the pioneer. The group individuals will have the option to try not to associate with the LBS worker. The pioneer will misuse the genuine questions with authentic situations as fakers to pick up full security insurance at his/her side. The LBS worker is mostly answerable for security approach execution, while the client's mission sends his/her question. This gathering expects that the LBS worker should be dependable. The security strategy executes on the cell phones of the LBS clients. The LBS worker views as a malignant segment. The putting away geographic guide alongside annoyance-based assurance settles the security issue on the client-side. It is a different network length. Hilbert bend planning changes over the two measurements of the put-away guide into one measure. In the third gathering, the client either settles on her security assurance choice dependent on the assistance given by the LBS worker side or associates with the LBS worker for the situation where no inquiry answer is in stock. The LBS worker is viewed as an assailant moreover. LBS worker reactions are stored to get profits by noting approaching questions with the advancement of time. The clients of every cell will be gathered in one group. A pioneer chooses a bunch. If a question-answer is not available in the stock, the backer will send the inquiry to the pioneer. The developer sends it to the untrusted LBS worker. The pioneer gets the appropriate response after controlling the question on the LBS worker side. It afterward restores the got answer to the needed client. The developer abuses genuine questions based on undisputed positions as fakers on his side. At that point, the Leader will send his question with the authentic situation with no compelling reason to deliver fakers either at the area or inquiry level. LBS worker reactions are stored to acquire profits by noting approaching inquiries with the advancement of time. In situations where no answer is available in the store, the LBS client compels to interface with the untrusted LBS supplier. Figure 1 portrays the system working.

Figure 1. The architecture of the system
(Alrahhal, Alrahhal, Jamous, & Jambi, 2020)

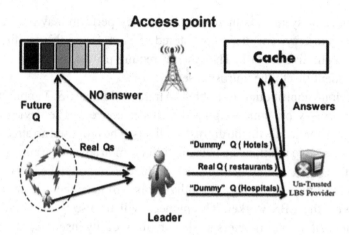

PROPOSED CONTRIBUTION

Table 1. Notations used in the work

Notations	Explanation
N	Network considered
H_i	Cluster head
S_i	Non-Trustworthy server
M_i	Cluster member
Id_H	Identity of the cluster head
Id_S	Identity of the server
Id_M	Identity of the cluster member
Q_i	Query generated by the user device
R_i	Requested data

The proposal adds reliability to the transmission by suffixing a hash message to the queries. The work uses the same foundation (Alrahhal, Alrahhal, Jamous, & Jambi, 2020). The activities include choosing the boundary for the devices and selecting the cluster head. Figure 2 depicts the system architecture

After the cluster head is chosen, the other nodes in the area share their hashed identities. In the area A_i, the cluster head H_i is sharing the identity with its cluster member M_i. The same is incorporated in the notation (1) and (2). In equation (1) the

Figure 2. System architecture of the proposed system

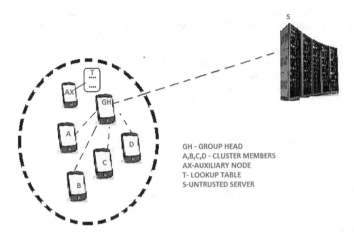

GH - GROUP HEAD
A,B,C,D - CLUSTER MEMBERS
AX-AUXILIARY NODE
T- LOOKUP TABLE
S-UNTRUSTED SERVER

hashed code of the cluster head identity id_H is transmitted by the cluster head H_i to the cluster member M_i. In equation (2) the hash code of the cluster member identity id_M is transmitted by the cluster member M_i to the cluster head H_i.

$$H_i \rightarrow M_i : hash\left(id_H\right) \qquad (1)$$

$$M_i \rightarrow H_i : hash\left(id_M\right) \qquad (2)$$

To send the queries, it generates the hash using its identity, location details, and time of dispatching. Let L_i be the location and T_i be the time of transmission. In Equation (3) the identity id_M, time of transmission T_i, location details L_i and query Q_i is transmitted to the cluster head H_i. This hashed code is detached from the query and forwarded to the untrusted server.

$$M_i \rightarrow H_i : Q_i \parallel hash\left(id_M, T_i, L_i\right) \qquad (3)$$

The similar methodology is used while the untrusted node S_i communicates with the cluster head H_i. It sends the requested data R_i along with the hashed code generated by its identity id_S, time of transmission T_j and location L_S. Equation (4) represents the same.

Table 2. Algorithm for hashing

Step 1: Input identification of the device (24 bits), time duration (20 bits), and location details (32 bits) (total – 76 bits) Step 2: concatenate all the bits one after another Step 3: Xor even position bits with odd position bits (resultant -38 bits) Step 4: Apply Right shift upto four positions. Step 5: Add even position bits with odd position bits (resultant – 20 bits)

$$S_i \rightarrow H_i : R_i \parallel hash\left(id_S, T_j, L_S\right) \tag{4}$$

Table 2 represents the algorithm used to generate the hash code.

Simulation

The work is simulated using NS2. Table 3 provides the setup used during simulation.

Table 3. Setup during simulation

Parameters Used	Explanation
Dimension of the network	200m * 200m
Number of nodes deployed	8 nodes
Number of clusters formed	2 cluster (4 nodes in each cluster)
Length of identification of the devices	24 bits
Length of location details	32 bits
Length of time of transmission	20 bits
Length of hash code	20 bits
Simulation time	60 s

ANALYSIS OF THE WORK

The proposal embeds the hashed code to the query to add reliability to the network. This methodology adds trust to the transmission.

- **Reliability** – The legitimate nodes communicating with one another build trust. The trust level decreases as the communicating party suspects some abnormal activity in the other node. If the devices are illegitimate, the receiving instrument will be able to find the guiltiness of the node after long time. Hence detecting the illegitimate node is essential. The suggestion enhances the trust of the communicating nodes. The communicating party suffixes the hashed code to the transmitting message before transmission. The proposal improves the reliability by 4.96% compared to the (Alrahhal, Alrahhal, Jamous, & Jambi, 2020). Figure 3 portrays the reliability of the proposal w.r.t the previous contribution.

Figure 3. Reliability of proposal w.r.t previous contribution

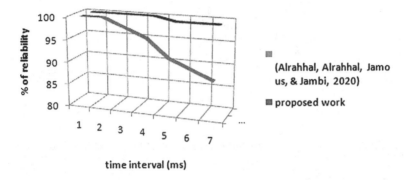

- **Early attack detection** – The receiving device will not be able to detect the illegitimacy of the node. This happens because the nodes are not programmed to do so. The proposal detects the guiltiness of the node at the early stage. As the hash code is affixed to the transmission and varies with time, the attacks can be detected. The time freshness is verified for attack detection. The location and device identity are unique to the device. Hence any illegitimacy can be detected. Figure 4 represents the detection of attack at an early stage by 1.31% compared to the previous work.

- **Energy conservation** – Energy is a rare resource that needs to be used with care. The devices loose energy if the instruments receive messages from illegitimate nodes. The device under attacks drains enormous energy with time. The proposal detects the attack at the early stage and hence conserves energy by 6.13%. Figure 5 represents the energy consumption in both the devices.

Figure 4. Early detection of attack

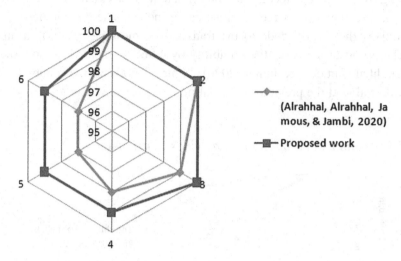

Figure 5. Energy conservation of the proposal w.r.t previous contribution

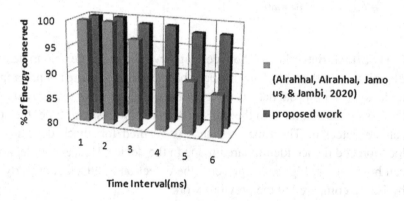

FUTURE WORK

The work focuses to maintain privacy of the users by enabling the heads to communicate with the illegitimate servers. The system can adopt future enhancement –

- Security is a primary concern in unsupervised environment. Methods have to adopt different kinds of encoding methodologies to secure data transmission. Some encryption standard can be adopted to enhance security in the network.
- The proposal tries detecting the attacks at an early stage by 1.31%. Better methodologies can be adopted to enhance the detection methods.

CONCLUSION

IoT are devices that are used in many applications. These devices are unsupervised and hence are liable to different kinds of attacks. The previous contribution implements privacy for its user. It adopts cluster-based implementation. The cluster head mediates between the untrusted server and user devices. The identity and their location are camouflaged in this work. The previous work aims in preserving privacy of the devices by camouflaging using cluster concept. To enhance the previous doing, the current proposal aims in securing the network, increasing reliability and conserving the energy of the network. The suggestion tries to implement some security measures to enhance reliability to the system. It suffixes the hash code derived from the identity of the device, location and transmission time. Apart from this it also preserves privacy in the network. The energy is conserved by 6.13%, the reliability is increased by 4.96% and the devices are able to detect attacks at the early stage by 1.31% compared to the previous contribution.

REFERENCES

Alpár, G. B., Batten, L., Moonsamy, V., Krasnova, A., Guellier, A., & Natgunanathan, I. (2016). New directions in IoT privacy using attribute-based authentication. In *ACM International Conference on Computing Frontiers* (pp. 461-466). Como Italy: ACM. 10.1145/2903150.2911710

Alrahhal, H., Alrahhal, M. S., Jamous, R., & Jambi, K. (2020). A Symbiotic Relationship Based Leader Approach for Privacy Protection in Location Based Services. *ISPRS International Journal of Geo-Information, 9*(6), 1–22. doi:10.3390/ijgi9060408

Ambika, N. (2020). Tackling Jamming Attacks in IoT. In *Internet of Things (IoT)* (pp. 153–165). Springer. doi:10.1007/978-3-030-37468-6_8

Ambika, N. (2021). TDSJ-IoT: Trivial Data Transmission to Sustain Energy From Reactive Jamming Attack in IoT. In Encyclopedia of Information Science and Technology, Fifth Edition (pp. 528-540). IGI Global.

Bahirat, P., He, Y., Menon, A., & Knijnenburg, B. (2018). A data-driven approach to developing IoT privacy-setting interfaces. In *23rd International Conference on Intelligent User Interfaces* (pp. 165-176). Tokyo Japan: ACM. 10.1145/3172944.3172982

Chanson, M., Bogner, A., Bilgeri, D., Fleisch, E., & Wortmann, F. (2019). Blockchain for the IoT: Privacy-preserving protection of sensor data. *Journal of the Association for Information Systems, 20*(9), 1274–1309. doi:10.17705/1jais.00567

Chen, S., Xu, H., Liu, D., Hu, B., & Wang, H. (2014). A vision of IoT: Applications, challenges, and opportunities with china perspective. *IEEE Internet of Things Journal, 1*(4), 349–359. doi:10.1109/JIOT.2014.2337336

Datta, T., Apthorpe, N., & Feamster, N. (2018). A developer-friendly library for smart home IoT privacy-preserving traffic obfuscation. In *Workshop on IoT Security and Privacy* (pp. 43-48). Budapest Hungary: ACM. 10.1145/3229565.3229567

Dorri, A., Kanhere, S. S., Jurdak, R., & Gauravaram, P. (2017). *Blockchain for IoT security and privacy: The case study of a smart home. In IEEE international conference on pervasive computing and communications workshops (PerCom workshops)*. IEEE.

Islam, M. N., & Kundu, S. (2018). Preserving IoT privacy in sharing economy via smart contract. In *IEEE/ACM Third International Conference on Internet-of-Things Design and Implementation (IoTDI)* (pp. 296-297). Orlando, FL: IEEE.

Junglas, I. A., & Watson, R. T. (2008). Location-based services. *Communications of the ACM, 51*(3), 65–69. doi:10.1145/1325555.1325568

Khan, M. A., & Salah, K. (2018). IoT security: Review, blockchain solutions, and open challenges. *Future Generation Computer Systems, 82*, 395–411. doi:10.1016/j.future.2017.11.022

Küpper, A. (2005). *Location-based services: fundamentals and operation*. John Wiley & Sons. doi:10.1002/0470092335

Naeini, P. E., Bhagavatula, S., Habib, H., Degeling, M., Bauer, L., Cranor, L. F., & Sadeh, N. (2017). Privacy expectations and preferences in an IoT world. In *Thirteenth Symposium on Usable Privacy and Security ({SOUPS})* (pp. 399-412). Santa Clara, CA: USENIX.

Nagaraj, A. (2021). Introduction to Sensors in IoT and Cloud Computing Applications. Bentham Science Publishers. doi:10.2174/97898114793591210101

Petrie, H., Johnson, V., Strothotte, T., Raab, A., Fritz, S., & Michel, R. (1996). MoBIC: Designing a travel aid for blind and elderly people. *Journal of Navigation, 49*(1), 45–52. doi:10.1017/S0373463300013084

Pokhrel, S. R., Vu, H. L., & Cricenti, A. L. (2019). Adaptive admission control for IoT applications in home WiFi networks. *IEEE Transactions on Mobile Computing*, *19*(12), 2731–2742. doi:10.1109/TMC.2019.2935719

Purri, S., Choudhury, T., Kashyap, N., & Kumar, P. (2017). Specialization of IoT applications in health care industries. In *International Conference on Big Data Analytics and Computational Intelligence (ICBDAC)* (pp. 252-256). Chirala, India: IEEE. 10.1109/ICBDACI.2017.8070843

Sun, G., Chang, V., Ramachandran, M., Sun, Z., Li, G., Yu, H., & Liao, D. (2017). Efficient location privacy algorithm for Internet of Things (IoT) services and applications. *Journal of Network and Computer Applications*, *89*, 3–13. doi:10.1016/j.jnca.2016.10.011

Sun, G., Liao, D., Li, H., Yu, H., & Chang, V. (2017). L2P2: A location-label based approach for privacy preserving in LBS. *Future Generation Computer Systems*, *74*, 375–384. doi:10.1016/j.future.2016.08.023

Tomatis, A., Cataldi, P., Pau, G., Mulassano, P., & Dovis, F. (2008). Cooperative LBS for Secure Transport System. *21st International Technical Meeting of the Satellite Division of The Institute of Navigation (ION GNSS 2008)*, 861-866.

Ukil, A., Bandyopadhyay, S., & Pal, A. (2014). IoT-privacy: To be private or not to be private. In *IEEE Conference on Computer Communications Workshops (INFOCOM WKSHPS)* (pp. 123-124). Toronto, Canada: IEEE. 10.1109/INFCOMW.2014.6849186

Yu, S. H. (2008). Methods for the Revitalization about LBS Mobile Games-Comparative Analysis between Internal and Overseas Case Study. *The Journal of the Korea Contents Association*, *8*(11), 74–84. doi:10.5392/JKCA.2008.8.11.074

Chapter 7

Internet of Things, Security of Data, and Cyber Security

Albérico Travassos Rosário

 https://orcid.org/0000-0003-4793-4110

GOVCOPP, IADE, Universidade Europeia, Portugal

ABSTRACT

Diverse forms of cyber security techniques are at the forefront of triggering digital security innovations, whereas cybersecurity has become one of the key areas of the internet of things (IoT). The IoT cybersecurity mitigates cybersecurity risk for organizations and users through tools such as blockchain, intelligent logistics, and smart home management. Literature has not provided the main streams of IoT cyber risk management trends, to cross referencing the diverse sectors involved of health, education, business, and energy, for example. This study aims to understanding the interplay between IoT cyber security and those distinct sector issues. It aims at identifying research trends in the field through a systematic bibliometric literature review (LRSB) of research on IoT cyber and security. The results were synthesized across current research subthemes. The results were synthesized across subthemes. The originality of the paper relies on its LRSB method, together with extant review of articles that have not been categorized so far. Implications for future research are suggested.

INTRODUCTION

The Internet of Things (IoT) refers to a system of interconnected and interrelated objects that can collect and transfer data over a wireless network. Ullah et al. (2019) defines it as the interconnection of physical moving objects, referred to as

DOI: 10.4018/978-1-7998-9312-7.ch007

"Things" embedded with sensors, electronic chips, and other types of hardware through the internet. Radio Frequency Identifier (RFID) tags are used to uniquely identify each device worldwide, allowing communication between the smart objects and the connected nodes (Ande et al., 2020). These features further enable remote monitoring and controlling. While IoT's rapid success has contributed to global digital transformation, the increased threats against IoT services and devices have not gone unnoticed (Abomhara & Køien, 2015). The interconnection and interrelation of the systems, services, applications, and data storage creates a gateway for cyberattacks, including malware attacks and software piracy threatening IoT security.

In the last few years, security attacks on the smart grid have significantly increased. Most system features in the current digital technologies depend on IoT to improve communication and operational efficiencies. Sani et al. (2019) indicate that the current security posture is inadequate to solve IoT threats and vulnerabilities. Thus, cyber security attacks have continued to cause environmental concerns and economic losses. Aich et al. (2019) explain that software and hardware technologies can be customized to function in different economic sectors. Consequently, digital transformation is occurring in all sectors of the economy, including education, automotive, pharmaceutical and healthcare, energy, and business. The lack of adequate strategies to fight against cyberattacks exposes these sectors to security threats that might undermine organizations and their customers' online safety.

Therefore, the purpose of this paper is to evaluate the correlation between IoT cyber security and distinct security issues in various sectors. A Systematic Bibliometric Literature Review (LRSB) will be conducted to identify and synthesis data on IoT cyber and security trends. The findings will contribute to a better understanding of the threats and attacks on IoT infrastructure and provide information that can be used to improve cyber defense.

METHODOLOGICAL APPROACH

Systematic Bibliometric Literature Review (LRSB) is a research methodology that enables researchers to offer comprehensive maps of the knowledge structure in various literature streams. Bibliometric reviews use statistical tools to identify trends, citations, and co-citations indicated in themes, authors, year, method, research problem, theory, journal, and country (Paul & Criado, 2020). In this research paper, the technique integrates systematic literature review techniques and bibliometric analysis to increase the accuracy of the literature analysis (Rosário, 2021, Raimundo & Rosário, 2021, Rosário et al., 2021, Rosário & Cruz, 2019). The process began with the search of a broad search query in the Scopus database, as it is recognized

Table 1. Process of systematic literature review.

Phase	Step	Description
Exploration	Step 1	formulating the research problem
	Step 2	searching for appropriate literature
	Step 3	critical appraisal of the selected studies
	Step 4	data synthesis from individual sources
Interpretation	Step 5	reporting findings and recommendations
Communicatio	Step 6	Presentation of the literature review report

Source: own elaboration

as one of the main indexes of scientific and/or academic documents reviewed by peer-reviewed.

The LRSB methodology used in this document has created valuable work providing vital information for business professionals and academics (Table 1).

The scientific document database used was Scopus, the most important peer review in the academic world. However, we consider that the study has the limitation of considering only the Scopus database, excluding other academic databases. The bibliographic search includes peer-reviewed scientific articles published until August 2021.

The keywords "Internet of Things" or "IoT" were used during the first search query to identify potential information sources. The search evaluated the titles, keywords, and abstract, and a total of 107,930 documents were identified.

A second search was limited to the subject area "Business, Management and Accounting", and the number of sources was reduced to 4,404.

The researcher determined that synthesizing data from all these sources would lead to irrelevant information that does not directly answer the research question, leading to the integration of more accurate keywords, including "Security Of Data," "Data Security," and "Cyber Security."

A total of 79 documents were selected through this screening process for analysis and data synthesis. Content and theme analysis techniques were used to identify, analyze and report the various studies as proposed by Rosário, 2021, Raimundo & Rosário, 2021, Rosário et al., 2021; Rosário & Cruz, 2019. The process is summarized in Table 2.

The 79 documents scientific are subsequently analyzed in a narrative manner to deepen the content and the possible derivation of common themes that directly answer the article's research question (Rosário, 2021, Raimundo & Rosário, 2021, Rosário et al., 2021; Rosário & Cruz, 2019). Of the 79 scientific documents selected, 50 Conference Paper, Article (25); Book Chapter (2); Book (1); and Review (1)..

Table 2. Screening methodology

Database Scopus	Screening	Publications
Meta-search	keyword: Internet of Things or IoT	107,930
Inclusion Criterion	keyword: Internet of Things or IoT Subject area Business, Management and Accounting	4,404
Screening	keyword: Internet of Things or IoT Subject area Business, Management and Accounting Exactkeyword: Security Of Data, or Data Security, Cyber Security	79

Source: own elaboration

PUBLICATION DISTRIBUTION

Peer-reviewed documents on the topic be period 2010-2021. The year 2020 were the one with the most peer-reviewed publications on the subject, reaching 20.

Figure 1 summarizes the peer-reviewed documents published and indexed in the Scopus database for the period 2010-2021.

The publications were sorted out as follows: Proceedings IEEE 2018 International Congress On Cybermatics 2018 IEEE Conferences On Internet Of Things Green Computing And Communications Cyber Physical And Social Computing Smart Data Blockchain Computer And Information Technology Ithings Greencom Cpscom Smartdata Blockchain CIT 2018 (10); Computer Law And Security Review (5); Proceedings Of The 2019 IEEE International Conference Quality Management Transport And Information Security Information Technologies IT And Qm And Is 2019 (4); with 2 (2019 IEEE Technology And Engineering Management Conference Temscon 2019; ACM Transactions On Management Information Systems; Annual Conference On Innovation And Technology In Computer Science Education Iticse; Big Data Management And Processing; Conference Proceedings Of The 7th International Symposium On Project Management Ispm 2019; IEEE Engineering Management Review; International Journal Of Recent Technology And Engineering; Journal Of Network And Systems Management; Lecture Notes In Business Information Processing; Proceedings Of The 2020 IEEE International Conference Quality Management Transport And Information Security Information Technologies IT And Qm And Is 2020) and with 1 (12th Aeit International Annual Conference Aeit 2020; 2011 International Conference On E Business And E Government Icee2011 Proceedings; 2016 International Conference On Information Technology Systems And Innovation Icitsi 2016 Proceedings; 2017 IEEE Technology And Engineering Management Society Conference Temscon 2017; 2018 IEEE International Conference On Engineering Technology And Innovation ICE Itmc 2018 Proceedings; 2018

IEEE Technology And Engineering Management Conference Temscon 2018; 2020 IEEE Technology And Engineering Management Conference Temscon 2020; 2020 International Conference On Computer Science Engineering And Applications Iccsea 2020; Advances In Production Engineering And Management; Advances In Transdisciplinary Engineering; Business Process Management Journal; How To Compete In The Age Of Artificial Intelligence Implementing A Collaborative Human Machine Strategy For Your Business;Icbc 2019 IEEE International Conference On Blockchain And Cryptocurrency; Iclem 2010 Logistics For Sustained Economic Development Infrastructure Information Integration Proceedings Of The 2010 International Conference Of Logistics Engineering And Management; Ictc 2019 10th International Conference On ICT Convergence ICT Convergence Leading The Autonomous Future; Idimt 2018 Strategic Modeling In Management Economy And Society 26th Interdisciplinary Information Management Talks; Information And Computer Security; International Conference On Management And Service Science Mass 2011; International Journal Of Automotive Technology And Management; International Journal Of Business Information Systems; International Journal Of Grid And Utility Computing; International Journal Of Information Management; International Journal Of Production Research; Journal Of Manufacturing Technology Management; Journal Of Telecommunications And The Digital Economy; Logforum; Proceedings 2019 IEEE 28th International Conference On Enabling Technologies Infrastructure For Collaborative Enterprises Wetice 2019; Proceedings 2019 IEEE 5th International Conference On Collaboration And Internet Computing Cic 2019; Proceedings 2020 IEEE International Conference On Blockchain Blockchain 2020; Proceedings 2021 21st Acis International Semi Virtual Winter Conference On Software Engineering Artificial Intelligence Networking And Parallel Distributed Computing Snpd Winter 2021; Proceedings Of 2013 IEEE International Conference On Service Operations And Logistics And Informatics Soli 2013; Proceedings Of International Conference On Intelligent Engineering And Management Iciem 2020; Proceedings Of International Conference On Research Innovation Knowledge Management And Technology Application For Business Sustainability Inbush 2020; Proceedings Of The 12th International Scientific And Technical Conference On Computer Sciences And Information Technologies Csit 2017; Proceedings Of The European Conference On Innovation And Entrepreneurship Ecie; Proceedings Of The International Conference On Industrial Engineering And Operations Management; Service Oriented Computing And Applications; Technological Forecasting And Social Change; Technology In Society; Wit Transactions On Information And Communication Technologies).

Interest in the subject has increased over time.

In Table 3 we analyze for the Scimago Journal & Country Rank (SJR), the best quartile and the H index by publication. Management Science is the most quoted

Figure 1. Documents by year
Source: own elaboration

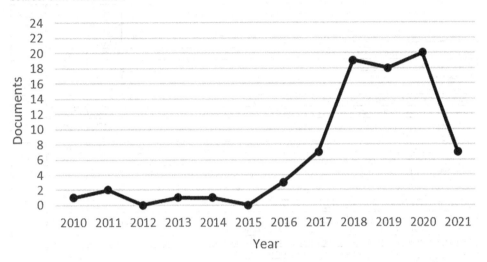

publication with 5,940 (SJR), Q1 and H index 127. There is a total of 7 publications on Q1, 3 publications on Q2 and 6 publications, Q3 and no publications in Q4. Publications from best quartile Q1 represent 14% of the 49 publications titles; best quartile Q2 represents 16%, best quartile and Q3 represents 12%, each of the titles of 49 publications.

Finally, 33 of the publications representing 67%, the data are not available.

As evident from Table 3, the significant majority of articles on of Internet of Things Security Of Data and Cyber Security rank on the Q3 best quartile index.

The subject areas covered by the 79 scientific documents were: Business, Management and Accounting (79); Computer Science (55); Decision Sciences (39); Engineering (32); Social Sciences (25); Medicine (11); Mathematics (6); Economics, Econometrics and Finance (3); Energy (2); Environmental Science (2); and with 1 (Physics and Astronomy; Psychology).

The most quoted publication was "Blockchain technology innovations" from Ahram et al. (2012) with 143 quotes published in the 2017 IEEE Technology and Engineering Management Society Conference, 0,210 (SJR), the best quartile (is not available) and with H index (6). The published publication is an effort to break the ground for presenting and demonstrating the use of Blockchain technology in multiple industrial applications.

In Figure 2 we can analyze the evolution of citations of publications published between ≤2010 and 2021. The number of quotes shows a positive net growth with an R2 of 65% for the period ≤2010-2021, with 2020 reaching 258 citations.

Table 3. Scimago journal and country rank impact factor.

Title	SJR	Best Quartile	H Index
International Journal Of Information Management	2,770	Q1	114
International Journal Of Production Research	1,910	Q1	142
Journal Of Manufacturing Technology Management	1,290	Q1	70
Computer Law And Security Review	0,820	Q1	38
Business Process Management Journal	0,670	Q1	81
Advances In Production Engineering And Management	0,620	Q1	18
ACM Transactions On Management Information Systems	0,600	Q1	29
Journal Of Network And Systems Management	0,490	Q2	35
International Journal Of Automotive Technology And Management	0,380	Q2	22
Information And Computer Security	0,330	Q2	49
IEEE Engineering Management Review	0,300	Q3	20
International Journal Of Business Information Systems	0,260	Q3	26
Lecture Notes In Business Information Processing	0,210	Q3	49
Journal Of Telecommunications And The Digital Economy	0,200	Q2	6
Logforum	0,200	Q3	4
International Journal Of Grid And Utility Computing	0,190	Q3	20
Icbc 2019 IEEE International Conference On Blockchain And Cryptocurrency	0,340	-*	8
Annual Conference On Innovation And Technology In Computer Science Education Iticse	0,260	-*	23
2018 IEEE International Conference On Engineering Technology And Innovation ICE Itmc 2018 Proceedings	0,220	-*	8
2017 IEEE Technology And Engineering Management Society Conference Temscon 2017	0,210	-*	6
Proceedings Of The 12th International Scientific And Technical Conference On Computer Sciences And Information Technologies Csit 2017	0,200	-*	13
2019 IEEE Technology And Engineering Management Conference Temscon 2019	0,150	-*	4
2016 International Conference On Information Technology Systems And Innovation Icitsi 2016 Proceedings	0,150	-*	6
Proceedings 2019 IEEE 28th International Conference On Enabling Technologies Infrastructure For Collaborative Enterprises Wetice 2019	0,130	-*	4
Proceedings Of The European Conference On Innovation And Entrepreneurship Ecie	0,130	-*	6
Proceedings Of The International Conference On Industrial Engineering And Operations Management	0,130	-*	9
2018 IEEE Technology And Engineering Management Conference Temscon 2018	0,120	-*	3
Ictc 2019 10th International Conference On ICT Convergence ICT Convergence Leading The Autonomous Future	0,120	-*	3
Conference Proceedings Of The 7th International Symposium On Project Management Ispm 2019	0,100	-*	1
Idimt 2018 Strategic Modeling In Management Economy And Society 26th Interdisciplinary Information Management Talks	0,100	-*	3
Proceedings Of 2013 IEEE International Conference On Service Operations And Logistics And Informatics Soli 2013	0	-*	7
International Conference On Management And Service Science Mass 2011	0	-*	6
Iclem 2010 Logistics For Sustained Economic Development Infrastructure Information Integration Proceedings Of The 2010 International Conference Of Logistics Engineering And Management	0	-*	5
International Journal Of Recent Technology And Engineering	-*	-*	20
Advances In Transdisciplinary Engineering	-*	-*	5
Proceedings IEEE 2018 International Congress On Cybermatics 2018 IEEE Conferences On Internet Of Things Green Computing And Communications Cyber Physical And Social Computing Smart Data Blockchain Computer And Information Technology Ithings Greencom Cpscom Smartdata Blockchain CIT 2018	-*	-*	-*
Proceedings Of The 2019 IEEE International Conference Quality Management Transport And Information Security Information Technologies IT And Qm And Is 2019	-*	-*	-*
Big Data Management And Processing	-*	-*	-*
Proceedings Of The 2020 IEEE International Conference Quality Management Transport And Information Security Information Technologies IT And Qm And Is 2020	-*	-*	-*
12th Aeit International Annual Conference Aeit 2020	-*	-*	-*
2011 International Conference On E Business And E Government Icee2011 Proceedings	-*	-*	-*
2020 IEEE Technology And Engineering Management Conference Temscon 2020	-*	-*	-*
2020 International Conference On Computer Science Engineering And Applications Iccsea 2020	-*	-*	-*
How To Compete In The Age Of Artificial Intelligence Implementing A Collaborative Human Machine Strategy For Your Business	-*	-*	-*
Proceedings 2019 IEEE 5th International Conference On Collaboration And Internet Computing Cic 2019	-*	-*	-*
Proceedings 2020 IEEE International Conference On Blockchain Blockchain 2020	-*	-*	-*
Proceedings 2021 21st Acis International Semi Virtual Winter Conference On Software Engineering Artificial Intelligence Networking And Parallel Distributed Computing Snpd Winter 2021	-*	-*	-*
Proceedings Of International Conference On Intelligent Engineering And Management Iciem 2020	-*	-*	-*
Proceedings Of International Conference On Research Innovation Knowledge Management And Technology Application For Business Sustainability Inbush 2020	-*	-*	-*

Note: *data not available.

Source: own elaboration

Figure 2. Evolution of citations between ≤2010 and 2021.
Source: own elaboration

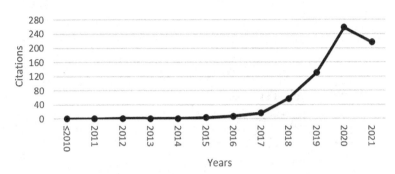

The h-index was used to ascertain the productivity and impact of the published work, based on the largest number of documents included that had at least the same number of citations. Of the documents considered for the h-index, 12 have been cited at least 12 times.

In Table 4 (Appendix), the citations of all scientific documents from the ≤2010 to 2021 period are analyzed, with a total of 691 citations, of the 79 publications 27 were not cited.

Table 5 examines the self-citation of the documents during the period ≤2010 to 2021, 24 documents were self-cited 65 times, the article A GDPR Controller for IoT Systems: Application to e-Health by Rhahla et al. (2019) Paper presented at the Proceedings - 2019 IEEE 28th International Conference on Enabling Technologies: Infrastructure for Collaborative Enterprises was cited 8 times.

In Figure 3, a bibliometric study was carried out to investigate and identify indicators on the dynamics and evolution of scientific information using the main keywords. The study of bibliometric results using the scientific software VOSviewe, aims at identifying the main research keywords "Internet of Things", "Security of Data", and "Cyber Security".

The research was based upon the studied documents on "internet of things security of data and cyber security". The linked keywords can be examined in Figure 4 making it possible to make clear the network of keywords that appear together / linked in each scientific documents, allowing to know the topics studied by the researches and to identify future research trends. In Figure 5, it is presented a profusion of co-citation with a unit of analysis of cited references.

Figure 3. Network of all keywords

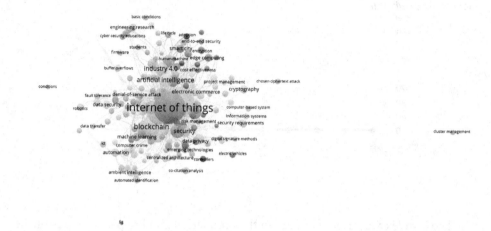

Figure 4. Network of linked keywords

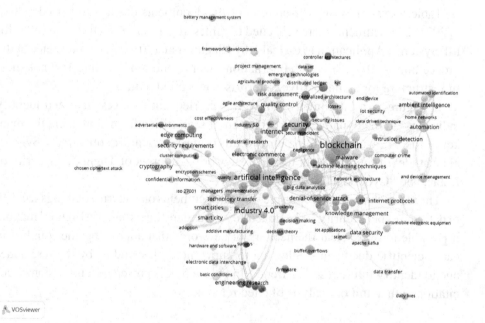

Figure 5. Network of co-citation

THEORETICAL PERSPECTIVES

The popularity of the Internet of Things (IoT) and its applications usher in a new era of computing. Every smart object is connected and equipped with features that allow data collection and communication over the internet. Consequently, IoT has been integrated into day-to-day activities through multiple applications of intelligent "things" (Saksonov et al., 2019). The various services and information offered through IoT environments apply in multiple fields, from healthcare to manufacturing (Asiri & Miri, 2018). While this integration has enhanced efficiency and performance, it has increased security threats for organizations using IoT systems and applications (Silverajan et al., 2018). Daim et al. (2020) indicate that cybersecurity in IoT occurs in three levels; organization, employee, and individual. IoT generates and assimilates massive amounts of data to facilitate business processes and influence society in general (Ahram et al., 2017). Thus, it collects organization and employees data to facilitate business decision-making and operational success, while per person data, such as consumer data, enables customer-centered products, systems, and procedures.

Internet of Things (IoT)

In the last few years, the popularity of the "Internet of Things" (IoT) has rapidly grown, with companies introducing numerous IoT-based products and services. For instance, the number of IoT acquisitions among tech companies has increased, e.g., Google acquired Nest for $3.2 billion, which later acquired Dropcam (Wortmann & Flüchter, 2015). At the same time, Samsung acquired SmartThings. These sophisticated technologies are believed to improve the quality of techniques and technologies used to collect, store, and share data (Sivakumar et al., 2019). However, Wortmann and Flüchter (2015) note that despite the popularity, there lacks a standard definition of IoT that describes what it encompasses. As a result, IoT has multiple meanings from varying researchers and developers. For instance, Caron et al. (2016) define disruptive technologies that integrate multiple communication and collaboration technologies that allow comprehensive data collection. Wortmann and Flüchter (2015) explain IoT as a global infrastructure for Information Society that virtually and physically interconnects things based on interoperable, evolving, and existing information and communication technologies. These features enable advanced services in numerous economic sectors.

IoT definitions vary due to the diversity of its elements. For instance, some definitions emphasize the interconnection capabilities of IoT and other internet-based features such as network technology and internet protocols (Caron et al., 2016). However, IoT can also be defined from different aspects, including the organization of large data volumes, search and storage capabilities (Liu et al., 2020). This diversity reflects on the increasingly extending IoT solutions to virtually all aspects of everyday activities. The main application area of IoT has been the smart industry, where the industrial internet of things (IIoT) contributes to the connected production sites and the intelligent production systems (Wortmann & Flüchter, 2015). Other sectors include smart transport systems, smart energy applications, smart homes and buildings, smart healthcare, smart cities, and intelligent security systems (Smith et al., 2021). IoT technologies in these industries are designed to monitor and provide appropriate real-time data to develop intelligent solutions (Saksonov et al., 2019). For instance, in smart health IoT technologies are used for chronic disease management and patient surveillance for improved care delivery and health outcomes. Healthcare professionals are interconnected with the appropriate IoT-based systems, increasing access to medical information and communication for evidence-based decision-making and interventions (Sohrabi Safa et al., 2017). The systems and processes digitalize functions and critical capabilities, creating significant opportunities for organizations and individuals.

Organizations and third parties use IoT to collect and analyze individual and environment characteristics to provide new reality-augmented and personalized services. For instance, a central cloud-based system gathers and analyzes domestic data for improved prediction on individual households' utility usage (Caron et al., 2016). While the monitoring and transmitting of the data through interconnected systems have increased organizational efficiency, it has also raised security and privacy concerns. Asiri and Miri (2018) explain that organizations' and third parties' demand for individual and personalized data has significantly grown in recent years, causing significant privacy and data safety concerns. In addition, the perceived value of individuals' and organizational data has resulted in increased targeted cyberattacks (Setiawan et al., 2017). For instance, hackers target financial institutions' cloud-based systems to acquire clients' financial information, including bank and credit or debit card details to conduct online frauds. The growing concerns over these online threats and thefts reflect the need to understand the correlation between IoT and cyber security.

Cyber Security

Advancements in internet and communication technologies and artificial intelligence have increased the number of distributed intelligent systems. As a result, the network environment and the categories and amounts of data flow handled in these environments have dramatically increased (Parasol, 2018). This expansion has made the spread of IoT and Cloud services essential as it allows people to transfer enormous data to cyberspace through heterogeneous devices (Rajashree et al., 2018). However, optimizing the opportunities created by this advancement requires protecting these devices and technologies to ensure the safety of the users' data and physical devices (Al-Omari et al., 2021). In this regard, effective cyber security measures are required to protect the systems, programs, and networks from digital attacks that aim to access, change or destroy sensitive information, disrupt business processes, or extort money from users. Gupta et al. (2019) define cyber security as a set of techniques and algorithms applied to protect the integrity of networks, nodes, and data from illegal access, damage, and attacks. Malicious access and threats against essential and sensitive information stored in cyberspace threaten the safety of individuals, businesses, and government institutions. An attack on IoT-based systems and Cloud services can interrupt business operations, cause mass panic, and destabilize the targeted environment.

Traditionally, cyber security has been treated as a technical issue, including defense techniques such as encryption and user authentication. However, the rapid development of intrusion techniques has prompted the expansion of the concept to include business and governance approaches (Dube & Mohanty, 2020). Executives are currently involved in assessing new threats and establishing effective response measures. Thus, cyber security is perceived as a critical component of risk management, accountability, and reporting (Almeida et al., 2020). Under the governance approach, the board and executive management exercise various responsibilities and practices to provide strategic direction (Parasol, 2018). This technique ensures that cyber security risks are appropriately managed and verifies responsible organizational resources to achieve specified security objectives. With information security (IS) governance, organizations perceive adequate security as a non-negotiable requirement, thus prioritizing defense infrastructure.

The primary goal of cyber security is to protect data and systems from malicious cyber threats that occur in different forms and areas. Examples of cyber threats include information foraging and applications outbreaks, malware, and viruses (Gupta et al., 2019). The rapid increase in cyberattacks has prompted organizations and enterprises to automate threat analysis as a measure of identifying vulnerabilities and threats before they cause significant damage (Puthal et al., 2017). Therefore, cyber security emphasizes developing defense tools, algorithms, and processes that unveil threat conditions and cyber attackers (Čapek, 2018). Al-Omari et al. (2021) recommend the use of Intrusion Detection Systems (IDS) that integrates Machine Learning (ML) as a defense tool. Alshboul et al. (2021) argue that ML can classify and predict attacks in cyberspace, allowing a timely response. Some ML-based techniques, such as Decision Trees, are used to create IDS based on classification algorithms used in supervised learning and can help identify anomalies and threats in cyberspace (Latif et al., 2021). However, technicians should look out for potential inaccuracies resulting from the complexity of the security features and the extensive network traffic data volumes.

IoT has increased vulnerabilities to cyberattacks. The most popular IoT application is the machine-to-machine (M2M) technology adopted in most industries, including oil, water, transportation, public service management, health, retail, and power (Abomhara & Køien, 2015). The interconnected has created a gateway for attackers, where illegal access to one machine can lead to a massive data breach within the centralized system (Beer & Hassan, 2018). For instance, in the current global supply chain, organizations depend on IT to communicate and coordinate collaboration between manufacturers, network suppliers, transport providers, and distributors (Latif et al., 2021). While this digitization increases visibility, information exchange, and agility, it exposes the organizations and contracted parties to potential cyberattacks (Oravec, 2017). This illustration reflects on the direct correlation between IoT and cyber security. To optimize IoT capabilities and opportunities, companies must prioritize cyber security and implement appropriate defense mechanisms.

Internet of Things (IoT) as an Enabling Technology for Industry 4.0

Industry 4.0 is a term used to refer to the fourth industrial revolution, characterized by smart manufacturing and digitization of business processes. It involves creating a connected ecosystem that combines physical production and operations with smart digital technologies, big data, and machine learning. Veile et al. (2020) defines Industry 4.0 as the connected and digitized industrial value creation characterized by connecting machines, people, ICT systems, and objects intelligently, vertically, and horizontally. Technological advancements yield multiple benefits and opportunities, including increased flexibility, efficiency, and quality (Soldani, 2020). Ardito et al. (2019) explain that the rapid development of Industry 4.0 is associated with organizations realizing the significance of information processing mechanisms as the fundamental component for digital transformation. IoT contributes to this by providing technologies that facilitate the gathering, analysis, storage, and sharing of data over the internet, fostering an integrated ecosystem for transparent collaborations.

The success of Industry 4.0 is dependent on the availability of relevant information to control and manage the entire value chain. It involves analyzing the generated data to optimize value flow through interconnected systems, people, and objects. Autenrieth et al. (2018) explain that Industry 4.0 combines the concepts of Cyber-Physical-Systems (CPS) and intelligent data analysis and interconnected systems. This description refers to the computational capabilities of the value chain objects, such as communication technologies and sensors, which are integrated into the IoT (Soltani et al., 2018). IoT is characterized by interoperability, enabling computerized systems or devices to readily connect and share information without restrictions (Dayarathna et al., 2017). In industry 4.0, these features support smart manufacturing processes through increased data acquisition and sharing. For instance, IoT-based systems use sensors' and systems' data from multiple machines within a company's system to provide predictive information (Autenrieth et al., 2018). Instead of relying on one machine, the company employees can integrate all the devices within the system to avoid downtimes. Therefore, the advanced computing and storage capabilities allowing the generation and analysis of large sets of data are essential for the growth of Industry 4.0.

Industry 4.0 includes cloud solutions that enhance employees' access to the required system features and data. The concept of Industry 4.0 is based on intelligent systems and technologies that can influence industrial and production processes (Voronova et al., 2019). According to Autenrieth et al. (2018), its IT infrastructure involves many software and hardware that require frequent maintenance and update. However, most employees need just a fraction of the features and not the entire IT infrastructure (Stephen Dass & Prabhu, 2020). Therefore, cloud solutions enable the implementation of virtualized and digitized computers characterized by customizable computer memory and power (De Carvalho & Eler, 2018). Employees install only the features required to carry out their tasks even though they access other components within the system if needed. The primary benefit of this technology is that no one has to administrate the system, and the IT infrastructure is highly scalable and available (Oconnor & Stricklan, 2021). In addition, Sarı et al. (2020) explain that IoT plays a critical role in facilitating the automation of various operations such as monitoring and controlling systems and processes. These processes can ease employees' burden of frequently updating and maintaining the functions and features needed for efficient workflow.

The primary aim of Industry 4.0 technologies is to improve value creation by interconnecting resources, people, information, and objects. The interconnection transforms the entire supply chain to increase efficiency, transparency, communication, and collaboration (Nekrasov & Polivoda, 2019). According to Veile et al. (2020), the interconnected systems created by IoT contribute to the new supply chains by enhancing real-time interconnection among supply chain players and customers. These key stakeholders benefit from IoT-based technologies through eased data exchange and analysis, facilitating informed decisions and planning (Terruggia & Garrone, 2020). For instance, supply chain partners can use the data gathered, analyzed, and shared to align supply chain processes, leading to improved resource efficiency (Stathaki et al., 2020). They can use the technologies to monitor and control energy consumption, material usage, and waste processes, thus, boosting production and reducing costs. Therefore, IoT as a fundamental technology of Industry 4.0 improves value creation processes by optimizing decision-making and enhancing supply chain flexibility.

Despite these opportunities, the digital transformation in Industry 4.0 faces multiple security challenges, including concerns for cyberattacks. The interconnectedness of smart manufacturing technologies attracts cyber attackers interested in various elements such as confidential production data or consumer information (Dhieb et al., 2020). In addition, information systems have always been perceived as liabilities, with the potential to cause significant damages (Culot et al., 2019). For instance, successful cyberattacks are often associated with employees' negligence, process failure, and malicious behaviors (Nash, 2021). Some workers can trade the organization's security elements such as codes and passwords, allowing attackers to intrude the systems. Large companies with large employee numbers have limited control over their behaviors and intentions. Therefore, open access to systems and data can lead to cyber vulnerabilities.

As a result, Managers associate cyber security with uncertainties and potential threats to business continuity, thus approaching the issue from cyber resilience and risk management perspectives. Various efforts are taken to control and monitor employees' activities, such as tracking the use of company resources such as machines (Dika & Nowostawski, 2018). For example, some IT executives can monitor teams' online activities to ensure they engage in only those aligned to the organization's business processes. In addition, some managers approach cyber security from the customer value creation perspective (Mohanty & Vyas, 2018). In this case, IoT-based technologies, big data, and cloud services are used to enhance data protection, provide reliable technologies, and strengthen secure access (Stathaki et al., 2020). Cyber security is prioritized as a means of competitive positioning by creating secure technologies and smart products. Customers are reluctant to engage in online business activities due to concerns over data protection and privacy (Foltz & Foltz, 2020). Therefore, prioritizing cyber security provides the necessary reassurance and builds trust. IoT, as part of digital transformation, provides customizable technologies that allow companies to achieve specific goals. Thus, integrating appropriate IoT-based technologies can contribute to cyber resilience, risk management, and competitive positioning.

Application of Internet of Things (IoT) in Healthcare Devices

IoT in healthcare has led to developments such as Ambient Assisted Living (AAL), remote patient monitoring, and smart health. It is thus, considered an attractive area for applying IoT to improve the quality of healthcare services (Mavropoulos et al., 2016). Healthcare practitioners use IoT-based healthcare applications for real-time and constant patient monitoring to develop progress reports on patient status (Khatkar et al., 2020). In this regard, combining IoT technologies and medical equipment allows early detection of threatening conditions and implementation of preventive interventions.

IoT-based healthcare systems play a critical role in connecting the available healthcare resources for improved care activities, including remote surgeries, diagnosis, and patient monitoring. It extends health care services from hospital settings to communities to homes through wireless technologies that connect monitoring devices (Rhahla et al., 2019). For instance, an IoT-based healthcare system connects patients to health care resources within the community, such as doctors, nurses, hospitals, ambulances, and rehabilitation centers (Khatkar et al., 2020). For instance, a rehabilitation center equipped with IoT's radio frequency identification tags (RFID) can analyze and consolidate data from a centralized database, identify critical events, and create rehabilitation strategies (Furstenau et al., 2020). Since all the objects are connected through the internet, the data collected and analyzed from multiple sources can be used to support an automated resource allocator to reduce health inequalities (Bhattacharya et al., 2020). Digital technologies can identify the ratio of resources availability versus population to identify gaps and recommend practical solutions (Martynov et al., 2019). In addition, IoT-based healthcare applications can be used to understand a group's specific needs, leading to customized care delivery (Yuehong et al., 2016). People's health needs vary due to multiple factors, individual characteristics, behaviors, social, economic, and physical environments. Understanding these determinants of health can facilitate targeted interventions and resources allocation in ways that address people's specific needs.

IoT can be used to control and monitor the use of resources within a facility for improved efficiency and allocation. According to Laplante and Laplante (2016), administration and insurance companies such as Medicare can use IoT to document the need for additional medical equipment through tracking and monitoring resources use within facilities. In addition, some healthcare staff tends to keep specific equipment, such as oxygen tanks and IV pumps, in their units for future use (Yuehong et al., 2016). Such actions can cause shortages for other departments in need of this equipment, especially in under-resourced institutions. Therefore, the hospital administration can use IoT-based technologies to monitor equipment use and ensure some practitioners are not undermining others' attempts to offer quality care. Besides, this approach can establish workplace policies that promote collaboration and equal access to resources for improved care delivery.

Other IoT applications in healthcare can include sensors, geolocation innovations, and smart security and surveillance. For instance, installing sensors in a patient's room can help detect body temperature or pressure (Laplante & Laplante, 2016). Similarly, biometric sensors can detect signs of aggressive behavior resulting from violence, which has become a severe issue in current healthcare environments (Giannoutakis et al., 2020). Geolocation innovations can be used to monitor Alzheimer's disease patients to limit unwanted movements or wandering. In contrast, smart security and surveillance can be used to securely and remotely monitor hospitals' IoT-based video surveillance technologies.

Despite these opportunities and benefits of applying IoT in healthcare, cyber security remains a significant challenge affecting its maximum integration. For instance, transferring data from IoT devices to cloud storage for easy access and transfer creates an opportunity for privacy and security breach (Bhattacharya et al., 2020). Healthcare data security is a sensitive issue highlighted in the Health Insurance Portability and Accountability Act of 1996 (HIPAA). HIPAA requires healthcare institutions to implement appropriate physical, administrative, and technical protection measures to ensure health data availability, confidentiality, and integrity. Sensitive patient information should be not be disclosed without consent or knowledge (Mangino et al., 2020). These requirements reflect the need to adopt defensive models that protect IoT health systems from potential attacks and breaches.

Internet of Things (IoT) and Cyber Security in Transport Sector

Modern vehicles evolve from mechanical systems to electromechanical systems that integrate software and hardware subsystems to create in-vehicle computer networks. IoT has significantly contributed to this transformation by changing the gathering of data and information in the transportation systems (Hu & Duan, 2019). IoT-based technologies bring together crucial business and technical aspects of automation, mobility, and data analytics (Gorog & Boult, 2018). The embedded actuators, sensors, and other IoT devices facilitate gathering and transmitting data on real-time activities with the transportation systems (Urquhart & McAuley, 2018). Consequently, this digital transformation in the sector has been associated with multiple opportunities, including:

Improved Road Safety

Connected vehicles have sensors that communicate with other cars, the cloud, and other IoT devices to collect data and information that can be used to improve road safety. The sensors and the cloud as the fundamental technologies of IoT facilitate data collection, analysis, and dispatch in a real-time and meaningful way (Hu & Duan, 2019). Examples of the data include weather and road conditions and vehicle maintenance issues, including tires' temperatures, battery state, fluid levels, and deterioration (Grandhi et al., 2021). This real-time information can be implemented to protect cars from potential road accidents. Morris et al. (2018) indicate that connected vehicles hold enormous amounts of personal data through carry-in devices and in-car networks. Drivers and fleet managers can use this data to make informed decisions focusing on improving the cars' state and promoting road safety.

Improved Operational Performance

IoT facilitates monitoring critical infrastructure, such as roads. Through the data gathered and shared, transportation agencies can understand the conditions of critical transport infrastructure and allocate appropriate resources for improvements (Griffy-Brown et al., 2018). In addition, key stakeholders can use the transmitted data to create more efficient processes to improve system capacity and reduce operating costs. According to Manoj Kumar and Dash (2017), executing smooth operations in the transportation system requires appropriate planning and effective transportation services. However, Mandrakov et al. (2020) indicate that the current sector is struggling with high costs, issues in navigation, inconveniences, lack of accountability, and service unreliability. Digital transformation based on IoT technologies provides digital solutions based on data and informational resources.

However, optimizing these opportunities requires proper IoT infrastructure for the transportation system. Manoj Kumar and Dash (2017, p.194) recommend that the adopted infrastructure have five layers: sensing, application, service, communication, and infrastructure. In the transportation system, IoT integrates information and communication technologies with automotive engineering design to facilitate digital transformation (Khalid et al., 2019). Given the increased adoption of technologies in all the other sectors of the economy and daily life activities, this sector needs to integrate appropriate IoT applications to match the advancements.

Application Layer

In an OSI model, the application layer is the closest to the end-user, allowing them to interact directly with the application software. This layer monitors different tasks

Figure 6. IoT architecture for transportation system
Source: Manoj Kumar & Dash (2017).

as required by clients (Griffy-Brown et al., 2019). In the transportation system, the functions monitored are associated with people, vehicles, terminals, traffic, service areas, goods, and junctions (Manoj Kumar & Dash, 2017). Other functions of this layer may include synchronizing communication, identifying partners, and evaluating the availability of resources.

Sensing Layer

The sensor layer contains a sensor network to facilitate interactions between vehicle captains the application layers through electronic devices. The sensors are integrated with the vehicle and include electronic devices such as RFIDs, cameras, infrared technology, sensors, image or text readers, and microwave technology (Uzunov et al., 2019). Examples of tasks carried out in this layer include detecting parking, collecting fees, monitoring the environment, tracking logistics, detecting passenger flow, monitoring compass terminals and cameras. These tasks ensure efficient workflow and performance of the transport system.

Communication Layer

The communication layer plays a significant role as the information bridge between the service and sensor layers. It is considered the backbone of the IoT system since it facilitates data transmission from the sensors to the service layer and other operating systems (Mohamed, 2019). The IoT devices installed in the transportation system collect enormous data that require transmitting to other connected systems and devices through a communication protocol (Ma & Wang, 2011). The transmission occurs over various internet connection methods including, Public and Private networks, WiFi, 3G/4G/5G Network, optical fiber, and wired network (Manoj Kumar & Dash, 2017). Other issues considered in this layer in the transportation system include the reliability and transparency of the data transfer and speed.

Service Layer

The service layer is responsible for performing the activities required by the client or the application player. It receives the information needed for processing from the sensing layer through the communication layer (Vasiliev & Aleksandrova, 2020).

Some components under the service layer identified by Manoj Kumar and Dash (2017, p.195) include Passenger Vehicle Platform, Highway Integrated Platform, Logistics Service Platform, Intelligent Travelling Service Platform, and Fleet Vehicle Service Platform. Various computing tools and analytics are used to processes the information and ensure that the performed activities align with the requirements

Infrastructure Layer

The infrastructure layer creates the technologies needed to perform the services within the transportation system. It includes a GIS mapping service, cloud storage, a cloud computing platform, and big data analytics tools (Manoj Kumar & Dash, 2017). Its various components facilitate system-to-system interactions, data storage, and retrieval to cater to requests and queries.

Despite the positive response to digital transformation in the transportation sector, cyber security remains a growing threat. For instance, Fiat Chrysler recalled 1.4 vehicles in July 2015 due to concerns over the cars' software and possible remote manipulation (Morris et al., 2018). Nissan Leaf was hacked through its NissanConnect EV application. The hack was associated with Software coding errors that granted hackers access to drivers' identity data and allowed remote control of the in-car systems (Wang et al., 2014). These cases reflected on potential vulnerabilities created by IoT-based technologies integrated into the transportation systems and connected vehicles. Therefore, it is essential to strengthening cyber security mechanisms to ensure the safety of all stakeholders. Although there are potential solutions, including software intrusion detection and source code security, hacking techniques are more complex to match the current technical advancements (Khalid et al., 2019). The hackers' knowledge and skills continue to grow with advancements in technology (Weber & Studer, 2016). Therefore, more research and developments are required to provide the appropriate technical solutions to cyber threats affecting the transportation sector and other fields adopting IoT-based technologies.

Internet of Things (IoT) in the Education Sector

Optimizing IoT in all sectors requires a workforce with the necessary knowledge and skills. Therefore, education institutions must develop and implement IoT courses and tools to ensure their graduates meet the current job market requirements. The existing educational institutions have changed from the knowledge transfer models to the active collaborative self-directed model facilitated by technologies (Bagheri & Movahed, 2016). The technologies have prompted educators to establish teaching models that focus on improving student outcomes through personalized content and higher student engagement (Foster et al., 2019). IoT plays a critical role in providing data needed to facilitate the transition. The primary goal of integrating IoT applications in the education system is to bridge the information gap that often undermines effective decision-making (Hitefield et al., 2018). For instance, student diversity in institutions hinders the implementation of customized learning and teaching techniques. The institutions and teachers can use data to understand their learners and create responsive models of learning based on their unique needs and characteristics.

IoT as enabling technology in higher education benefits the institutions and students. For instance, while the institutions can depend on enormous data gathered and transmitted to control the inventory and track the movement of resources, students can use the technologies to access educational materials (Ibrahim et al., 2020). QR codes and embedded sensors can help students explore educational environments, such as the library, and access learning materials from anywhere at any given time (Bagheri & Movahed, 2016). The various applications of IoT within these learning environments have created 'smart classroom,' which refers to an intelligent learning place equipped with different hardware and software to enhance student outcomes (Foster et al., 2019). Some technologies found in smart classrooms include face recognition algorithms, sensors, cameras, and video projectors. They can collect students' personal attributes data, such as achievements, performance, ability to focus, and physical environment data.

Examples of IoT applications in higher education include:

Creating a Safe and Secure Classroom and Campus Access Control

Universities experience challenges associated with managing students' access to facilities such as laboratories and classrooms safely and securely. Integrating Radio Frequency Identification (RFID) and Near Field Communication (NFC) can help address these problems (Bagheri & Movahed, 2016). For instance, the classroom registration system can be based on NFC technologies to create a real-time classroom control tool. The connected sensors collect classroom access information, then displayed it on the university TV panels or a web-based application (Kis & Singh, 2018). Similarly, RFID tags can be integrated into student ID cards to monitor campus attendance and location based on geofencing technology (Webster et al., 2018). The installation of IoT infrastructure can create a smart environment that protects students and staff wellbeing by simplifying access to safe spaces (Leang et al., 2018). For instance, during the COVID-19 period, IoT monitoring can be used to ensure that the number of students accessing a specific classroom or other school resources observes the required physical distancing guidelines. For example, a mobile application can be adapted to allow students to check classroom capacity, reserve appropriate seats, and availability of free desks in the library.

Improving Learning and Teaching

IoT provides education institutions with real-time actionable insight about learners' performance and creates a richer learning experience, thus improving the quality of teaching and learning. IoT-based technologies can be used to develop smart learning environments characterized by customizable features that match student needs (Wolff & Nuseibah, 2017). For example, a smart learning environment can allow learners to adjust physical environment variables such as room temperature, noise, and CO2 levels to maximize their capability to focus on the lectures and other educational activities. These changes can significantly impact students' attitudes, mood, and perception of the institution, increasing the probability of achieving higher academic goals. In addition, IoT can be used to track and monitor individual student characteristics such as study patterns through devices such as virtual and augmented reality headsets, cameras, and sensors (Bagheri & Movahed, 2016). Other resources such as academic databases can be used to enhance knowledge, while classroom data can be implemented to create personalized teaching techniques.

Monitoring Students' Health

Access to quality health care services is essential in any academic establishment since students' health status directly impacts their academic performance. IoT-based healthcare applications facilitate access to quality health care, reduce care costs, help monitor patients' health, and prevent diseases through early identification of potential health threats (Rhahla et al., 2019). Wearable technologies such as smartwatches and fitness bands are the most common IoT applications in health monitoring (Lee et al., 2019). They can be used to monitor student health conditions such as blood pressure and other vital signs. In addition, IoT promotes eHealth by providing and storing students' health information such as medical histories, laboratory tests, and prescriptions. This data can be accessed remotely through connected devices, ensuring that health practitioners working with the student are adequately informed about their vital signs, health conditions, and treatments.

CONCLUSION

The Internet of Things (IoT) refers to the interconnection and interrelation of smart objects through a wireless network. IoT devices such as sensors, electronic chips, and RFID tags are used to collect, analyze and transmit data to connected systems and devices for further action. In addition, IoT-based technologies enable remote controlling and monitoring of the connected objects to facilitate digital transformation. These advanced technologies enable firms to create a centralized platform that enhances collaboration and communication. These opportunities increase IoT's applicability in multiple fields, including Industry 4.0, healthcare, transportation, and education.

Industry 4.0 involves smart manufacturing, an ecosystem of interconnected systems, and smart data analysis technologies. In healthcare, IoT provides smart devices such as smartwatches that can monitor a patient's health condition and vital signs such as body temperature and pressure. The shift to electromechanical systems has increased the need for in-vehicle computer networks based on software and hardware subsystems. This integration has improved road safety and operational performance through enhanced technologies that prioritize stakeholders' needs. Optimizing these opportunities requires necessary computing knowledge and skills. Therefore, educational institutions are responding accordingly by integrating IoT courses and tools in their systems. As a result, the institutions have improved the quality of teaching and learning, leading to higher academic achievements. Thus, IoT technologies can be integrated into diverse sectors to improve productivity, performance, efficiency, and convenience.

However, the interconnectivity in IoT creates a gateway for cyber-attacks, indicate the need to prioritize cybersecurity to safeguard sensitive information and critical systems. Unlike in the past, where cybersecurity was considered a technical issue, the current digital environment requires the integration of business and governance approaches. The prevalence of cybersecurity concerns in all sectors indicates the dire need to strengthen defense mechanisms to avoid the potential damages caused by cybersecurity threats.

ACKNOWLEDGMENT

We would like to express our gratitude to the Editor and the Referees. They offered extremely valuable suggestions or improvements. The authors were supported by the GOVCOPP Research Center of Universidade de Aveiro and Universidade Europeia.

REFERENCES

Abomhara, M., & Køien, G. M. (2015). Cyber security and the internet of things: Vulnerabilities, threats, intruders and attacks. *Journal of Cyber Security and Mobility*, 65-88.

Ahram, T., Sargolzaei, A., Sargolzaei, S., Daniels, J., & Amaba, B. (2017). Blockchain technology innovations. *IEEE Technology and Engineering Management Society Conference, TEMSCON 2017*, 137-141. 10.1109/TEMSCON.2017.7998367

Aich, S., Chakraborty, S., Sain, M., Lee, H. I., & Kim, H. C. (2019, February). A review on benefits of IoT integrated blockchain based supply chain management implementations across different sectors with case study. In *2019 21st international conference on advanced communication technology (ICACT)* (pp. 138-141). IEEE. 10.23919/ICACT.2019.8701910

Al-Omari, M., Rawashdeh, M., Qutaishat, F., Alshira'H, M., & Ababneh, N. (2021). An intelligent tree-based intrusion detection model for cyber security. *Journal of Network and Systems Management*, 29(2), 20. Advance online publication. doi:10.100710922-021-09591-y

Almeida, F., Duarte Santos, J., & Augusto Monteiro, J. (2020). The challenges and opportunities in the digitalization of companies in a post-COVID-19 world. *IEEE Engineering Management Review*, 48(3), 97–103. doi:10.1109/EMR.2020.3013206

Alshboul, Y., Bsoul, A. A. R., & Zamil, A. L. (2021). Cybersecurity of smart home systems: Sensor identity protection. *Journal of Network and Systems Management*, 29(3), 22. Advance online publication. doi:10.100710922-021-09586-9

Ande, R., Adebisi, B., Hammoudeh, M., & Saleem, J. (2020). Internet of Things: Evolution and technologies from a security perspective. *Sustainable Cities and Society*, 54, 101728. doi:10.1016/j.scs.2019.101728

Ardito, L., Petruzzelli, A. M., Panniello, U., & Garavelli, A. C. (2019). Towards industry 4.0: Mapping digital technologies for supply chain management-marketing integration. *Business Process Management Journal*, 25(2), 323–346. doi:10.1108/BPMJ-04-2017-0088

Asiri, S., & Miri, A. (2018). A sybil resistant IoT trust model using blockchains. In *IEEE Conferences on Internet of Things, Green Computing and Communications, Cyber, Physical and Social Computing, Smart Data, Blockchain, Computer and Information Technology, iThings/GreenCom/CPSCom/SmartData/Blockchain/CIT 2018*, 1017-1026. 10.1109/Cybermatics_2018.2018.00190

Autenrieth, P., Lorcher, C., Pfeiffer, C., Winkens, T., & Martin, L. (2018). Current significance of IT-infrastructure enabling industry 4.0 in large companies. *2018 IEEE International Conference on Engineering, Technology and Innovation, ICE/ITMC 2018 – Proceedings*. 10.1109/ICE.2018.8436244

Bagheri, M., & Movahed, S. H. (2016, November). The effect of the Internet of Things (IoT) on education business model. In *2016 12th International Conference on Signal-Image Technology & Internet-Based Systems (SITIS)* (pp. 435-441). IEEE.

Beer, M. I., & Hassan, M. F. (2018). Adaptive security architecture for protecting RESTful web services in enterprise computing environment. *Service Oriented Computing and Applications, 12*(2), 111–121. doi:10.100711761-017-0221-1

Bhattacharya, S., Senapati, S., Soy, S. K., Misra, C., & Barik, R. K. (2020). Performance analysis of enhanced mist-assisted cloud computing model for healthcare system. *2020 International Conference on Computer Science, Engineering and Applications, ICCSEA 2020*. 10.1109/ICCSEA49143.2020.9132914

Čapek, J. (2018). Cybersecurity and internet of things. *IDIMT 2018: Strategic Modeling in Management, Economy and Society - 26th Interdisciplinary Information Management Talks,* 343-349.

Caron, X., Bosua, R., Maynard, S. B., & Ahmad, A. (2016). The internet of things (IoT) and its impact on individual privacy: An australian perspective. *Computer Law & Security Review, 32*(1), 4–15. doi:10.1016/j.clsr.2015.12.001

Culot, G., Fattori, F., Podrecca, M., & Sartor, M. (2019). Addressing industry 4.0 cybersecurity challenges. *IEEE Engineering Management Review, 47*(3), 79–86. doi:10.1109/EMR.2019.2927559

Daim, T., Lai, K. K., Yalcin, H., Alsoubie, F., & Kumar, V. (2020). Forecasting technological positioning through technology knowledge redundancy: Patent citation analysis of IoT, cybersecurity, and blockchain. *Technological Forecasting and Social Change, 161*, 120329. Advance online publication. doi:10.1016/j.techfore.2020.120329

Dayarathna, M., Fremantle, P., Perera, S., & Suhothayan, S. (2017). Role of real-time big data processing in the internet of things. In Big data management and processing (pp. 239-262) doi:10.1201/9781315154008

De Carvalho, L. G., & Eler, M. M. (2018). *Security requirements and tests for smart toys.* doi:10.1007/978-3-319-93375-7_14

Dhieb, N., Ghazzai, H., Besbes, H., & Massoud, Y. (2020). Scalable and secure architecture for distributed IoT systems. *2020 IEEE Technology and Engineering Management Conference, TEMSCON 2020.* 10.1109/TEMSCON47658.2020.9140108

Dika, A., & Nowostawski, M. (2018). Security vulnerabilities in ethereum smart contracts. *Proceedings - IEEE 2018 International Congress on Cybermatics: 2018 IEEE Conferences on Internet of Things, Green Computing and Communications, Cyber, Physical and Social Computing, Smart Data, Blockchain, Computer and Information Technology, iThings/GreenCom/CPSCom/SmartData/Blockchain/CIT 2018,* 955-962. 10.1109/Cybermatics_2018.2018.00182

Dube, D. P., & Mohanty, R. P. (2020). Towards development of a cyber-security capability maturity model. *International Journal of Business Information Systems,* *34*(1), 104–127. doi:10.1504/IJBIS.2020.106800

Foltz, C. B., & Foltz, L. (2020). Mobile users' information privacy concerns instrument and IoT. *Information and Computer Security, 28*(3), 359–371. doi:10.1108/ICS-07-2019-0090

Foster, D., White, L., Erdil, D. C., Adams, J., Argüelles, A., Hainey, B., ... Stott, L. (2019). Toward a cloud computing learning community. *Annual Conference on Innovation and Technology in Computer Science Education, ITiCSE,* 143-155. 10.1145/3344429.3372506

Furstenau, L. B., Sott, M. K., Homrich, A. J. O., Kipper, L. M., Al Abri, A. A., Cardoso, T. F., ... Cobo, M. J. (2020). 20 years of scientific evolution of cyber security: A science mapping. *Proceedings of the International Conference on Industrial Engineering and Operations Management,* 314-325.

Giannoutakis, K. M., Spathoulas, G., Filelis-Papadopoulos, C. K., Collen, A., Anagnostopoulos, M., Votis, K., & Nijdam, N. A. (2020). A blockchain solution for enhancing cybersecurity defence of IoT. *IEEE International Conference on Blockchain, Blockchain 2020,* 490-495. 10.1109/Blockchain50366.2020.00071

Gorog, C., & Boult, T. E. (2018). Solving global cybersecurity problems by connecting trust using blockchain. *Proceedings - IEEE 2018 International Congress on Cybermatics: 2018 IEEE Conferences on Internet of Things, Green Computing and Communications, Cyber, Physical and Social Computing, Smart Data, Blockchain, Computer and Information Technology, iThings/GreenCom/CPSCom/SmartData/Blockchain/CIT 2018,* 1425-1432. 10.1109/Cybermatics_2018.2018.00243

Grandhi, L. S., Grandhi, S., & Wibowo, S. (2021). A security-UTAUT framework for evaluating key security determinants in smart city adoption by the Australian city councils. *Proceedings - 2021 21st ACIS International Semi-Virtual Winter Conference on Software Engineering, Artificial Intelligence, Networking and Parallel/Distributed Computing,* 17-22. 10.1109/SNPDWinter52325.2021.00013

Griffy-Brown, C., Lazarikos, D., & Chun, M. (2018). Agile business growth and cyber risk. *2018 IEEE Technology and Engineering Management Conference, TEMSCON 2018.* 10.1109/TEMSCON.2018.8488397

Griffy-Brown, C., Miller, H., Zhao, V., Lazarikos, D., & Chun, M. (2019). Emerging technologies and risk: How do we optimize enterprise risk when deploying emerging technologies? *2019 IEEE Technology and Engineering Management Conference, TEMSCON 2019.* 10.1109/TEMSCON.2019.8813743

Gupta, S., Sabitha, A. S., & Punhani, R. (2019). Cyber security threat intelligence using data mining techniques and artificial intelligence. *International Journal of Recent Technology and Engineering, 8*(3), 6133–6140. doi:10.35940/ijrte.C5675.098319

Hitefield, S. D., Fowler, M., & Clancy, T. C. (2018). Exploiting buffer overflow vulnerabilities in software defined radios. *Proceedings - IEEE 2018 International Congress on Cybermatics: 2018 IEEE Conferences on Internet of Things, Green Computing and Communications, Cyber, Physical and Social Computing, Smart Data, Blockchain, Computer and Information Technology, iThings/GreenCom/CPSCom/SmartData/Blockchain/CIT 2018,* 1921-1927. 10.1109/Cybermatics_2018.2018.00318

Hu, G., & Duan, X. (2019). Research on the measures of safety supervision in road transportation of dangerous goods. *Conference Proceedings of the 7th International Symposium on Project Management, ISPM 2019,* 874-879.

Ibrahim, S., Shukla, V. K., & Bathla, R. (2020). Security enhancement in smart home management through multimodal biometric and passcode. *Proceedings of International Conference on Intelligent Engineering and Management, ICIEM 2020,* 420-424. 10.1109/ICIEM48762.2020.9160331

Khalid, A., Sundararajan, A., Hernandez, A., & Sarwat, A. I. (2019). FACTS approach to address cybersecurity issues in electric vehicle battery systems. *2019 IEEE Technology and Engineering Management Conference, TEMSCON 2019,* 10.1109/TEMSCON.2019.8813669

Khatkar, M., Kumar, K., & Kumar, B. (2020). An overview of distributed denial of service and internet of things in healthcare devices. *Proceedings of International Conference on Research, Innovation, Knowledge Management and Technology Application for Business Sustainability, INBUSH 2020*, 44-48. 10.1109/INBUSH46973.2020.9392171

Kis, M., & Singh, B. (2018). A cybersecurity case for the adoption of blockchain in the financial industry. *Proceedings - IEEE 2018 International Congress on Cybermatics: 2018 IEEE Conferences on Internet of Things, Green Computing and Communications, Cyber, Physical and Social Computing, Smart Data, Blockchain, Computer and Information Technology, Things/GreenCom/CPSCom/SmartData/Blockchain/CIT 2018*, 1491-1498. 10.1109/Cybermatics_2018.2018.00252

Laplante, P. A., & Laplante, N. (2016). The internet of things in healthcare: Potential applications and challenges. *IT Professional, 18*(3), 2–4. doi:10.1109/MITP.2016.42

Latif, M. N. A., Aziz, N. A. A., Hussin, N. S. N., & Aziz, Z. A. (2021). Cyber security in supply chain management: A systematic review. *Logforum, 17*(1), 49–57. doi:10.17270/J.LOG.2021555

Leang, B., Kim, R., & Yoo, K. (2018). Real-time transmission of secured plcs sensing data. *Proceedings - IEEE 2018 International Congress on Cybermatics: 2018 IEEE Conferences on Internet of Things, Green Computing and Communications, Cyber, Physical and Social Computing, Smart Data, Blockchain, Computer and Information Technology, iThings/GreenCom/CPSCom/SmartData/Blockchain/CIT 2018*, 931-932. 10.1109/Cybermatics_2018.2018.00177

Lee, T., Kim, S., & Kim, K. (2019). A research on the vulnerabilities of PLC using search engine. *ICTC 2019 - 10th International Conference on ICT Convergence: ICT Convergence Leading the Autonomous Future*, 184-188. 10.1109/ICTC46691.2019.8939961

Liu, C., Feng, Y., Lin, D., Wu, L., & Guo, M. (2020). Iot based laundry services: An application of big data analytics, intelligent logistics management, and machine learning techniques. *International Journal of Production Research, 58*(17), 5113–5131. doi:10.1080/00207543.2019.1677961

Ma, L., & Wang, Y. (2011). Research on formation mechanism of information risk in supply chain and its control countermeasures. *International Conference on Management and Service Science, MASS 2011*, 10.1109/ICMSS.2011.5998265

Mandrakov, E. S., Vasiliev, V. A., & Dudina, D. A. (2020). Non-conforming products management in a digital quality management system. *Proceedings of the 2020 IEEE International Conference "Quality Management, Transport and Information Security, Information Technologies", IT and QM and IS 2020*, 266-268. 10.1109/ITQMIS51053.2020.9322931

Mangino, A., Pour, M. S., & Bou-Harb, E. (2020). Internet-scale insecurity of consumer internet of things. *ACM Transactions on Management Information Systems*, *11*(4), 1–24. Advance online publication. doi:10.1145/3394504

Manoj Kumar, N., & Dash, A. (2017, November). Internet of things: an opportunity for transportation and logistics. *Proceedings of the International Conference on Inventive Computing and Informatics (ICICI 2017)*, 194-197.

Martynov, V. V., Shavaleeva, D. N., & Zaytseva, A. A. (2019). Information technology as the basis for transformation into a digital society and industry 5.0. *Proceedings of the 2019 IEEE International Conference Quality Management, Transport and Information Security, Information Technologies IT and QM and IS 2019*, 539-543. 10.1109/ITQMIS.2019.8928305

Mavropoulos, O., Mouratidis, H., Fish, A., Panaousis, E., & Kalloniatis, C. (2016). *APPARATUS: Reasoning about security requirements in the internet of things*. doi:10.1007/978-3-319-39564-7_21

Mohamed, K. S. (2019). IoT Networking and Communication Layer. In *The Era of Internet of Things* (pp. 49–70). Springer. doi:10.1007/978-3-030-18133-8_3

Mohanty, S., & Vyas, S. (2018). How to compete in the age of artificial intelligence: Implementing a collaborative human-machine strategy for your business. How to compete in the age of artificial intelligence: Implementing a collaborative human-machine strategy for your business. doi:10.1007/978-1-4842-3808-0

Morris, D., Madzudzo, G., & Garcia-Perez, A. (2018). Cybersecurity and the auto industry: The growing challenges presented by connected cars. *International Journal of Automotive Technology and Management*, *18*(2), 105–118. doi:10.1504/IJATM.2018.092187

Nash, I. (2021). Cybersecurity in a post-data environment: Considerations on the regulation of code and the role of producer and consumer liability in smart devices. *Computer Law & Security Review*, *40*, 105529. Advance online publication. doi:10.1016/j.clsr.2021.105529

Nekrasov, H. A., & Polivoda, D. E. (2019). The development of a reference architectural prototype of the internet of things network mode. *Proceedings of the 2019 IEEE International Conference Quality Management, Transport and Information Security, Information Technologies IT and QM and IS 2019*, 554-557. 10.1109/ITQMIS.2019.8928303

Oconnor, T. J., & Stricklan, C. (2021). Teaching a hands-on mobile and wireless cybersecurity course. *Annual Conference on Innovation and Technology in Computer Science Education, ITiCSE*, 296-302. 10.1145/3430665.3456346

Oravec, J. A. (2017). Kill switches, remote deletion, and intelligent agents: Framing everyday household cybersecurity in the internet of things. *Technology in Society*, *51*, 189–198. doi:10.1016/j.techsoc.2017.09.004

Parasol, M. (2018). The impact of china's 2016 cyber security law on foreign technology firms, and on china's big data and smart city dreams. *Computer Law & Security Review*, *34*(1), 67–98. doi:10.1016/j.clsr.2017.05.022

Paul, J., & Criado, A. R. (2020). The art of writing literature review: What do we know and what do we need to know? *International Business Review*, *29*(4), 101717. doi:10.1016/j.ibusrev.2020.101717

Puthal, D., Nepal, S., Ranjan, R., & Chen, J. (2017). End-to-end security framework for big sensing data streams. Big Data Management and Processing, 263-278. doi:10.1201/9781315154008

Raimundo, R., & Rosário, A. (2021). Blockchain system in the Higher Education. *European Journal of Investigation in Health, Psychology and Education, 11*(1), 276-293. doi:10.3390/ejihpe1101002

Rajashree, S., Gajkumar Shah, P., & Murali, S. (2018). Security model for internet of things end devices. *Proceedings - IEEE 2018 International Congress on Cybermatics: 2018 IEEE Conferences on Internet of Things, Green Computing and Communications, Cyber, Physical and Social Computing, Smart Data, Blockchain, Computer and Information Technology, iThings/GreenCom/CPSCom/SmartData/Blockchain/CIT 2018*, 219-221. 10.1109/Cybermatics_2018.2018.00066

Rhahla, M., Abdellatif, T., Attia, R., & Berrayana, W. (2019). A GDPR controller for IoT systems: Application to e-health. *Proceedings - 2019 IEEE 28th International Conference on Enabling Technologies: Infrastructure for Collaborative Enterprises, WETICE 2019,* 170-173. 10.1109/WETICE.2019.00044

Rosário, A. (2021). Research-Based Guidelines for Marketing Information Systems. *International Journal of Business Strategy and Automation*, 2(1), 1–16. doi:10.4018/IJBSA.20210101.oa1

Rosário, A., & Cruz, R. (2019). Determinants of Innovation in Digital Marketing, Innovation Policy and Trends in the Digital Age. *Journal of Reviews on Global Economics*, 8, 1722–1731. doi:10.6000/1929-7092.2019.08.154

Rosário, A., Fernandes, F., Raimundo, R., & Cruz, R. (2021). Determinants of Nascent Entrepreneurship Development. In A. Carrizo Moreira & J. G. Dantas (Eds.), *Handbook of Research on Nascent Entrepreneurship and Creating New Ventures* (pp. 172–193). IGI Global. doi:10.4018/978-1-7998-4826-4.ch008

Saksonov, E. A., Leokhin, Y. L., & Azarov, V. N. (2019). Organization of information security in industrial internet of things systems. *2019 IEEE International Conference Quality Management, Transport and Information Security, Information Technologies IT and QM and IS 2019*, 3-7. 10.1109/ITQMIS.2019.8928442

Sani, A. S., Yuan, D., Jin, J., Gao, L., Yu, S., & Dong, Z. Y. (2019). Cyber security framework for Internet of Things-based Energy Internet. *Future Generation Computer Systems*, 93, 849–859. doi:10.1016/j.future.2018.01.029

Sarı, T., Güleş, H. K., & Yiğitol, B. (2020). Awareness and readiness of industry 4.0: The case of Turkish manufacturing industry. *Advances in Production Engineering & Management*, 15(1), 57–68. doi:10.14743/apem2020.1.349

Setiawan, A. B., Syamsudin, A., & Sastrosubroto, A. S. (2017). Information security governance on national cyber physical systems. *2016 International Conference on Information Technology Systems and Innovation, ICITSI 2016 - Proceedings*, 10.1109/ICITSI.2016.7858210

Silverajan, B., Ocak, M., & Nagel, B. (2018). Cybersecurity attacks and defences for unmanned smart ships. *2018 International Congress on Cybermatics: 2018 IEEE Conferences on Internet of Things, Green Computing and Communications, Cyber, Physical and Social Computing, Smart Data, Blockchain, Computer and Information Technology, iThings/GreenCom/CPSCom/SmartData/Blockchain/CIT 2018*, 15-20. 10.1109/Cybermatics_2018.2018.00037

Sivakumar, S., Siddappa Naidu, K., & Karunanithi, K. (2019). Design of energy management system using autonomous hybrid micro-grid under IOT environment. *International Journal of Recent Technology and Engineering*, 8(2), 338-343. doi:10.35940/ijrte.B1058.0782S219

Smith, K. J., Dhillon, G., & Carter, L. (2021). User values and the development of a cybersecurity public policy for the IoT. *International Journal of Information Management, 56,* 102123. Advance online publication. doi:10.1016/j.ijinfomgt.2020.102123

Sohrabi Safa, N., Maple, C., & Watson, T. (2017). An information security risk management model for smart industries. *Advances in Transdisciplinary Engineering, 6,* 257-262. 10.3233/978-1-61499-792-4-257

Soldani, D. (2020). On Australia's cyber and critical technology international engagement strategy towards 6G how Australia may become a leader in cyberspace. *Journal of Telecommunications and the Digital Economy, 8*(4), 127–158. doi:10.18080/jtde.v8n4.340

Soltani, R., Nguyen, U. T., & An, A. (2018). A new approach to client onboarding using self-sovereign identity and distributed ledger. *Proceedings - IEEE 2018 International Congress on Cybermatics: 2018 IEEE Conferences on Internet of Things, Green Computing and Communications, Cyber, Physical and Social Computing, Smart Data, Blockchain, Computer and Information Technology, iThings/GreenCom/CPSCom/SmartData/Blockchain/CIT 2018,* 1129-1136. 10.1109/Cybermatics_2018.2018.00205

Stathaki, C., Xenakis, A., Skayannis, P., & Stamoulis, G. (2020). Studying the role of proximity in advancing innovation partnerships at the dawn of industry 4.0 era. *Proceedings of the European Conference on Innovation and Entrepreneurship, ECIE,* 651-658. doi:10.34190/EIE.20.048

Stephen Dass, A., & Prabhu, J. (2020). Hybrid coherent encryption scheme for multimedia big data management using cryptographic encryption methods. *International Journal of Grid and Utility Computing, 11*(4), 496–508. doi:10.1504/IJGUC.2020.108449

Terruggia, R., & Garrone, F. (2020). Secure IoT and cloud based infrastructure for the monitoring of power consumption and asset control. *12th AEIT International Annual Conference, AEIT 2020.* 10.23919/AEIT50178.2020.9241195

Ullah, F., Naeem, H., Jabbar, S., Khalid, S., Latif, M. A., Al-Turjman, F., & Mostarda, L. (2019). Cyber security threats detection in internet of things using deep learning approach. *IEEE Access: Practical Innovations, Open Solutions, 7,* 124379–124389. doi:10.1109/ACCESS.2019.2937347

Urquhart, L., & McAuley, D. (2018). Avoiding the internet of insecure industrial things. *Computer Law & Security Review, 34*(3), 450–466. doi:10.1016/j.clsr.2017.12.004

Uzunov, A. V., Nepal, S., & Baruwal Chhetri, M. (2019). Proactive antifragility: A new paradigm for next-generation cyber defence at the edge. *Proceedings - 2019 IEEE 5th International Conference on Collaboration and Internet Computing, CIC 2019*, 246-255. 10.1109/CIC48465.2019.00039

Vasiliev, V. A., & Aleksandrova, S. V. (2020). The prospects for the creation of a digital quality management system DQMS. *Proceedings of the 2020 IEEE International Conference "Quality Management, Transport and Information Security, Information Technologies", IT and QM and IS 2020*, 53-55. 10.1109/ITQMIS51053.2020.9322890

Veile, J. W., Kiel, D., Müller, J. M., & Voigt, K. (2020). Lessons learned from industry 4.0 implementation in the German manufacturing industry. *Journal of Manufacturing Technology Management, 31*(5), 977–997. doi:10.1108/JMTM-08-2018-0270

Voronova, L. I., Bezumnov, D. N., & Voronov, V. I. (2019). Development of the research stand «smart city systems» INDUSTRY 4.0. *Proceedings of the 2019 IEEE International Conference Quality Management, Transport and Information Security, Information Technologies IT and QM and IS 2019*, 577-582. 10.1109/ITQMIS.2019.8928370

Wang, F., Lin, T., Tsai, H., & Lu, Y. (2014). Applying RSA signature scheme to enhance information security for RFID based power meter system. *WIT Transactions on Information and Communication Technologies, 49*, 549-556. 10.2495/ICIE130652

Weber, R. H., & Studer, E. (2016). Cybersecurity in the internet of things: Legal aspects. *Computer Law & Security Review, 32*(5), 715–728. doi:10.1016/j.clsr.2016.07.002

Webster, G. D., Harris, R. L., Hanif, Z. D., Hembree, B. A., Grossklags, J., & Eckert, C. (2018). Sharing is caring: Collaborative analysis and real-time enquiry for security analytics. *Proceedings - IEEE 2018 International Congress on Cybermatics: 2018 IEEE Conferences on Internet of Things, Green Computing and Communications, Cyber, Physical and Social Computing, Smart Data, Blockchain, Computer and Information Technology, iThings/GreenCom/CPSCom/SmartData/Blockchain/CIT 2018*, 1402-1409. 10.1109/Cybermatics_2018.2018.00240

Wolff, C., & Nuseibah, A. (2017). A projectized path towards an effective industry-university-cluster: Ruhrvalley. *Proceedings of the 12th International Scientific and Technical Conference on Computer Sciences and Information Technologies, CSIT 2017, 2*, 123-131. 10.1109/STC-CSIT.2017.8099437

Wortmann, F., & Flüchter, K. (2015). Internet of things. *Business & Information Systems Engineering, 57*(3), 221–224. doi:10.100712599-015-0383-3

Yuehong, Y. I. N., Zeng, Y., Chen, X., & Fan, Y. (2016). The internet of things in healthcare: An overview. *Journal of Industrial Information Integration*, *1*, 3–13. doi:10.1016/j.jii.2016.03.004

KEY TERMS AND DEFINITIONS

Ambient Assisted Living: A subcategory of environmental intelligence, using environmental intelligence techniques, processes, and technologies.

Cyber Security: Cybersecurity is the protection of computer systems against theft or damage to electronic hardware, software, or data.

Industry 4.0: Fourth industrial revolution, technologies for automation and data exchange that uses concepts of cyber-physical systems, internet of things, and cloud computing.

Internet of Things: A concept that refers to the digital interconnection of everyday objects with the internet.

Intrusion Detection Systems: Also known as Intrusion Detection System refers to the technical means of discovering unauthorized access in a network.

Security of Data: Data protection of a database, that is, protection from unwanted actions by unauthorized users, such as cyber attack or data breach.

APPENDIX

Table 4. Overview of document citations period ≤2010 to 2021

Documents		≤2010	2011	2012	2013	2014	2015	2016	2017	2018	2019	2020	2021	Total
An Intelligent Tree-Based Intrusion Detection Model for Cybe...	2021	-	-	-	-	-	-	-	-	-	-	-	1	1
User values and the development of a cybersecurity public po...	2021	-	-	-	-	-	-	-	-	-	-	1	1	2
Internet-scale Insecurity of Consumer Internet of Things	2020	-	-	-	-	-	-	-	-	-	-	1	-	1
Forecasting technological positioning through technology kno...	2020	-	-	-	-	-	-	-	-	-	-	-	7	7
Lessons learned from Industry . implementation in the Germ...	2020	-	-	-	-	-	-	-	-	-	5	13	26	44
A Blockchain Solution for Enhancing Cybersecurity Defence of...	2020	-	-	-	-	-	-	-	-	-	-	-	1	1
Iot based laundry services: an application of big data analy...	2020	-	-	-	-	-	-	-	-	-	-	5	11	16
Mobile users' information privacy concerns instrument and Io...	2020	-	-	-	-	-	-	-	-	-	-	1	1	2
The Challenges and Opportunities in the Digitalization of Co...	2020	-	-	-	-	-	-	-	-	-	-	-	12	12
Performance Analysis of Enhanced Mist-Assisted Cloud Computi...	2020	-	-	-	-	-	-	-	-	-	-	-	1	1
Awareness and readiness of Industry .: The case of Turkish...	2020	-	-	-	-	-	-	-	-	-	-	1	5	6
years of scienti c evolution of cyber security: A scienc...	2020	-	-	-	-	-	-	-	-	-	-	11	4	15
Toward a cloud computing learning community	2019	-	-	-	-	-	-	-	-	-	-	1	2	3
Proactive antifragility: A new paradigm for next-generation ...	2019	-	-	-	-	-	-	-	-	-	-	-	1	1
A Research on the Vulnerabilities of PLC using Search Engine	2019	-	-	-	-	-	-	-	-	-	-	1	-	1
Organization of Information Security in Industrial Internet ...	2019	-	-	-	-	-	-	-	-	-	-	2	1	3
Information Technology as the Basis for Transformation into ...	2019	-	-	-	-	-	-	-	-	-	-	4	4	8
Development of the Research Stand «smart City Systems» INDUS...	2019	-	-	-	-	-	-	-	-	-	-	1	1	2

continued on following page

Table 4. Continued

Documents		≤2010	2011	2012	2013	2014	2015	2016	2017	2018	2019	2020	2021	Total
Cyber security threat intelligence using data mining techniq...	2019	-	-	-	-	-	-	-	-	-	-	2	1	**3**
Addressing Industry . Cybersecurity Challenges	2019	-	-	-	-	-	-	-	-	-	1	9	10	**20**
A GDPR Controller for IoT Systems: Application to e-Health	2019	-	-	-	-	-	-	-	-	-	1	6	2	**9**
FACTS approach to address cybersecurity issues in electric v...	2019	-	-	-	-	-	-	-	-	-	4	3	2	**9**
BlockONS: Blockchain based Object Name Service	2019	-	-	-	-	-	-	-	-	-	-	4	2	**6**
Towards Industry .: Mapping digital technologies for suppl...	2019	-	-	-	-	-	-	-	-	-	12	48	38	**99**
Flexible, e cient, and secure access delegation in cloud c...	2019	-	-	-	-	-	-	-	-	-	-	3	4	**7**
Agile Business Growth and Cyber Risk:	2018	-	-	-	-	-	-	-	-	-	1	1	-	**2**
How to compete in the age of arti cial intelligence: Implem...	2018	-	-	-	-	-	-	-	-	-	1	1	1	**3**
Current Signi cance of IT-Infrastructure Enabling Industry ...	2018	-	-	-	-	-	-	-	-	-	1	1	2	**4**
Security Vulnerabilities in Ethereum Smart Contracts	2018	-	-	-	-	-	-	-	-	-	1	4	3	**8**
A New Approach to Client Onboarding Using Self-Sovereign Ide...	2018	-	-	-	-	-	-	-	-	-	-	6	2	**8**
Solving Global Cybersecurity Problems by Connecting Trust Us...	2018	-	-	-	-	-	-	-	-	-	1	-	-	**1**
A Cybersecurity Case for the Adoption of Blockchain in the F...	2018	-	-	-	-	-	-	-	-	-	-	1	-	**1**
Sharing is Caring: Collaborative Analysis and Real-Time Enqu...	2018	-	-	-	-	-	-	-	-	-	-	2	-	**2**
Real-Time transmission of secured plcs sensing data	2018	-	-	-	-	-	-	-	-	-	-	-	1	**1**
Cybersecurity Attacks and Defences for Unmanned Smart Ships	2018	-	-	-	-	-	-	-	-	-	-	5	-	**5**
Security Model for Internet of Things End Devices	2018	-	-	-	-	-	-	-	-	-	1	1	1	**3**
A Sybil Resistant IoT Trust Model Using Blockchains	2018	-	-	-	-	-	-	-	-	-	-	4	-	**4**
Avoiding the internet of insecure industrial things	2018	-	-	-	-	-	-	-	-	4	14	7	7	**32**
Adaptive security architecture for protecting RESTful web se...	2018	-	-	-	-	-	-	-	-	-	1	3	3	**7**
The impact of China's Cyber Security Law on foreign tec...	2018	-	-	-	-	-	-	-	-	6	3	6	4	**19**

continued on following page

183

Table 4. Continued

Documents		≤2010	2011	2012	2013	2014	2015	2016	2017	2018	2019	2020	2021	Total
Security requirements and tests for smart toys	2018	-	-	-	-	-	-	-	-	-	-	3	-	3
Cybersecurity and the auto industry: The growing challenges ...	2018	-	-	-	-	-	-	-	-	-	3	5	1	9
A projectized path towards an e ective industry-university-...	2017	-	-	-	-	-	-	-	-	-	1	2	-	3
Kill switches, remote deletion, and intelligent agents: Fram...	2017	-	-	-	-	-	-	-	-	2	-	2	2	6
Blockchain technology innovations	2017	-	-	-	-	-	-	-	-	11	45	59	28	143
Information security governance on national cyber physical s...	2017	-	-	-	-	-	-	-	-	1	1	-	-	2
An information security risk management model for smart indu...	2017	-	-	-	-	-	-	-	-	-	1	1	1	3
Cybersecurity in the Internet of Things: Legal aspects	2016	-	-	-	-	-	-	-	4	16	14	9	7	50
The Internet of Things (IoT) and its impact on individual pr...	2016	-	-	-	-	-	-	4	7	12	9	15	12	59
APPARATUS: Reasoning about security requirements in the inte...	2016	-	-	-	-	-	-	-	5	1	2	1	-	9
Study on intelligent port under the construction of smart ci...	2013	-	-	-	-	-	-	1	-	1	2	1	1	6
Evaluation on security system of internet of things based on...	2011			1	1	1	3	2	-	3	5	1	2	19
		0	0	1	1	1	3	7	16	57	130	258	216	691

Source: own elaboration

Table 5. Overview of document self-citation period ≤2010 to 2021

Documents		≤2010	2011	2012	2013	2014	2015	2016	2017	2018	2019	2020	2021	Total
Forecasting technological positioning through technology kno...	2020	-	-	-	-	-	-	-	-	-	-	-	1	**1**
Lessons learned from Industry . implementation in the Germ...	2020	-	-	-	-	-	-	-	-	-	-	1	1	**3**
Iot based laundry services: an application of big data analy...	2020	-	-	-	-	-	-	-	-	-	-	-	1	**1**
The Challenges and Opportunities in the Digitalization of Co...	2020	-	-	-	-	-	-	-	-	-	-	-	4	**4**
20 years of scienti c evolution of cyber security: A scienc...	2020	-	-	-	-	-	-	-	-	-	-	3	-	**3**
Toward a cloud computing learning community	2019	-	-	-	-	-	-	-	-	-	-	1	2	**3**
Organization of Information Security in Industrial Internet ...	2019	-	-	-	-	-	-	-	-	-	-	1	1	**2**
Information Technology as the Basis for Transformation into ...	2019	-	-	-	-	-	-	-	-	-	-	3	3	**6**
Development of the Research Stand «smart City Systems» INDUS...	2019	-	-	-	-	-	-	-	-	-	-	1	2	**3**
Cyber security threat intelligence using data mining techniq...	2019	-	-	-	-	-	-	-	-	-	-	2	1	**3**
Emerging technologies and risk: How do we optimize enterpris...	2019	-	-	-	-	-	-	-	-	-	-	-	1	**1**
A GDPR Controller for IoT Systems: Application to e-Health	2019	-	-	-	-	-	-	-	-	-	4	3	1	**8**
FACTS approach to address cybersecurity issues in electric v...	2019	-	-	-	-	-	-	-	-	-	1	-	-	**1**
BlockONS: Blockchain based Object Name Service	2019	-	-	-	-	-	-	-	-	-	1	1	-	**2**
Flexible, e cient, and secure access delegation in cloud c...	2019	-	-	-	-	-	-	-	-	-	-	2	3	**5**
Avoiding the internet of insecure industrial things	2018	-	-	-	-	-	-	-	-	-	2	-	-	**2**
Adaptive security architecture for protecting RESTful web se...	2018	-	-	-	-	-	-	-	-	-	-	-	1	**1**
Security requirements and tests for smart toys	2018	-	-	-	-	-	-	-	-	-	-	2	-	**2**
Cybersecurity and the auto industry: The growing challenges ...	2018	-	-	-	-	-	-	-	-	-	-	1	-	**1**
A projectized path towards an e ective industry-university-...	2017	-	-	-	-	-	-	-	-	-	-	1	-	**1**
Blockchain technology innovations	2017	-	-	-	-	-	-	-	-	1	1	-	1	**3**
Cybersecurity in the Internet of Things: Legal aspects	2016	-	-	-	-	-	-	1	-	1	-	-		**2**
The Internet of Things (IoT) and its impact on individual pr...	2016	-	-	-	-	-	1	1	-	-	-	-	1	**3**
APPARATUS: Reasoning about security requirements in the inte...	2016	-	-	-	-	-		3	-	1	-	-		**4**
		0	**0**	**0**	**0**	**0**	**0**	**1**	**5**	**1**	**11**	**22**	**24**	**65**

Source: own elaboration

Chapter 8
An Effective Secured Privacy–Protecting Data Aggregation Method in IoT

Sabyasachi Pramanik
https://orcid.org/0000-0002-9431-8751
Haldia Institute of Technology, India

ABSTRACT

Because privacy concerns in IoT devices are the most sensitive of all the difficulties, such an extreme growth in IoT usage has an impact on the privacy and life spans of IoT devices, because until now, all devices communicated one to one, resulting in high traffic that may shorten the life of unit nodes. In addition, delivering data repeatedly increases the likelihood of an attacker attacking the system. Such traffic may exacerbate security concerns. The employment of an aggregator in the system as an intermediary between end nodes and the sink may overcome these problems. In any system with numerous sensors or nodes and a common controller or sink, we can use an intermediate device to combine all of the individual sensor data and deliver it to the sink in a single packet. Aggregator is the name given to such a device or component. Data aggregation is carried out to decrease traffic or communication overhead. In general, this strategy helps to extend the life of a node while also reducing network transmission.

INTRODUCTION

This chapter briefly discusses the differences between IoT privacy and security concerns, as well as several IoT privacy-preserving approaches like as anonymization,

DOI: 10.4018/978-1-7998-9312-7.ch008

dummies, PIR (Sahoo, C. K. et. al., 2017), caching (Siddiqui, I. F. 2020), collaborating, and so on. Location, encryption (Pramanik, S. et. al., 2020), and homomorphic approaches are all types of anonymization strategies for privacy (Pramanik, S. et. al., 2014) protection. The authors also go through each layer of the internet of things' privacy protection strategy in depth.

Security and Privacy

As prior debates in this area show, a number of researchers have viewed privacy as a component of security. Despite numerous large overlaps and a few crossovers, there is significant variation between these two. These differences are depicted here. Individual individuals with their distinct personal/sensitive data are referred to as "privacy" which emphasises the necessity of not exploiting the grounds for data protection, unlawful data access, or improper data usage that is not in accordance with the user's wishes. Furthermore, privacy may be defined as the degree to which each person interacts with the environment, as well as the amount of data that is authorised for public viewing. The phrase "security" refers to the endeavour to secure data and devices against outside threats, spyware, and subversion. Manufacturers often place a higher priority on hardware security (Sarkar, M. et. al. 2020, Pramanik, S. et. al. 2014) than on user privacy (Pramanik, S. et. al., 2019, Pandey, B. K. et. al., 2022). In particular, privacy invasion is achievable by collecting user data and then analysing it, while in terms of security, just obtaining information such as a user's password (Pramanik, S. et. al., 2017) is sufficient. Furthermore, security offers data protection inside trusted-parties, which largely includes foreign assault; but, in the case of privacy, there are certain un-trusted parties, such as Service Providers (SP), who must be dealt with within it. This study focuses on one area in particular. In an IoT context, privacy is important. Fig. 1 illustrates the fundamental distinctions between privacy and security.

There are currently no techniques that concentrate on data privacy preservation in a device, but data preservation through links/servers with varying degrees of confidence in SP, Third Party (TP), peers, and no complete trust has been the major emphasis thus far. In many IoT applications, such as smart devices (Pramanik, S. et. al. 2021), social media, phone apps, LBS, smart healthcare (Dutta, S. et. al. 2021), and so on, there are numerous cases of privacy breaches.

BACKGROUND

As shown in Fig. 1, the IoT privacy-protecting techniques are categorised into the following 11 groups based on the characteristics of prior privacy-preserving

Figure 1. The distinction between privacy and security.

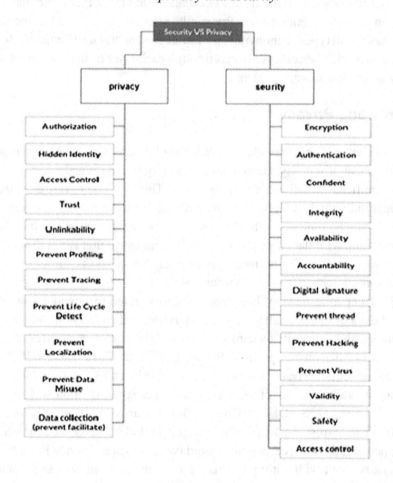

solutions in the IoT (Kaur, M. J., 2020; Sinha, M. et. al, 2021). These are further divided into four classes based on the trust problem. Every class has its own set of advantages and disadvantages, as well as certain unresolved issues and problems. There is a special approach-based technique someplace, and when this is the case, it has unique behaviour that distinguishes it from others.

MAIN FOCUS OF THE CHAPTER

Various techniques for privacy preservation in IoT are described below:

Figure 2. Techniques of privacy classification

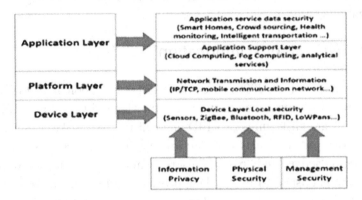

Anonymity

Individual records within a record group are rendered identical using generalisation, suppression, and/or clustering (Mandal, A. et. al. 2021) methods in order to protect individual privacy. For disclosing the concealed identity, this strategy will be more advantageous if alternative queries are considered and identity relationships as well as relationships among different sensitive data (location, time, etc.) are kept secret. Sweeney presented the k-anonymity model, which was the first anonymization model ever suggested, and it has been used since m-invariance, l-diversity, customised anonymity, and t-closeness are examples of advanced anonymization models.

Pros:

- All parties must respect each other's privacy.
- There's no need to put your faith in anybody.
- It's simple to use.
- It's useful for securing any form of privacy.

Cons:

- When a user connects to the internet using SP, it is inactive, and so the user's privacy is breached.
- In the Internet of Things, there is an anonymization technique.
- Among the several strategies available to assure privacy protection in IoT (Sinha, M. et. al. 2021), anonymization is regarded the most fundamental. The authors have two sorts of privacy protection in the IoT.

This section discusses the analysis used to accomplish these two goals using the anonymization approach.

Protection of Personal Information

RFID (Elbasani, E. et. al. 2020) is seen as a critical component of the Internet of Things. Low prices, mass manufacture, and the flexibility to attach RFID tags to possible items are all features of RFID-enabled IoT that make it omnipresent. It will raise severe privacy issues since the IoT may infiltrate people's daily lives and influence both "active" and "passive" users. If unique identification of persons is feasible in the published RFID tag data, private information will be exposed, and uniquely identifying information will be removed to prevent identification.

However, this modification does not ensure that individuals' personal information is kept private in the data. Anonymization ensures that no individual's identity is disclosed. Sweeny's k-anonymity model offers a more realistic technique for privacy-related issues preservation, which has piqued the curiosity of numerous academics. Here, a number of efficient algorithms are used. This is guaranteed by Sweeny's k-anonymity model, which states that each record is similar to a minimum of (k-1) other records in terms of quasi-identifier properties. As a result, retrieving any privacy-related information from a k-anonymity protected database is impossible, although it results in significant data loss, affecting data processing accuracy.

Privacy Protection in the Workplace

In the same manner that anonymization protects data privacy; it may also protect the location privacy of IoT nodes. Location-based services are an essential use of the Internet of Things (LBS). When a location server gets user requests for a location service like GPS (Pradhan, D. et. al., 2022) services, one of the primary issues with IoT is safeguarding the privacy of specific locations. Anonymization may be used to effectively safeguard individual location privacy information. The following are the specific procedures used in this case:

1. Individual information may be anonymized by allowing trusted anonymous third party admission into the LBS and user.
2. If the LBS server (Talaat, F. M., 2020) needs to be queried, location data is provided to a trusted anonymous third party.
3. Sent location data is a designated region with a large number of other users in order to safeguard an individual's practical location. Because all users must trust the anonymous third party, this strategy limits the degree of privacy protection, resulting in a single point attack.

IoT Encryption Technique

With the fast rise of IoT, huge opportunities for data processing have opened up, such as data mining, data query, and so on. Data processing activities are carried out in groups based on the private inputs provided by the participants. All of these duties might take place in the presence of mutually untrustworthy parties/competitors. As a result, the key concern presented here is the protection of personal information. To address this issue in a distributed setting, two commonly used encryption approaches are outlined here:

- SMC (Secure Multi-Party Computing) (Baouya A., et. al. 2021) is a system that allows for secure multi-party computation.
- Encryption that is homomorphic (Shankar, K. et. al. 2020)

SMC

In algorithms for distributed privacy-preserving data processing, cooperation among participants is required for result calculation; otherwise, no-sensitive processing results would be shared, potentially revealing sensitive information. The Millionaires' Problem of Nakai (Nakai, T. 2021) leads to the formation of SMC. The primary issue here is that two billionaires want to know who is the wealthiest among them without divulging their net worth. The issue entails a simple comparison of two numbers; one from each side, with neither party informed the other of its number. Secure comparison, secure set union, dot product protocol, secure sum, secure intersection, and other secure sub-protocols are often employed in this technology.

IoT Homomorphic Encryption

A fog orchestration paradigm is developed to control both reaction times and service delivery issues caused by diverse IoT security methods. By enabling the network to be tailored, this technology provides desired services, including crucial privacy and security solutions. In addition, attribute-based encryption and homomorphic encryption are used to secure data privacy in IoT devices while reducing reaction time and power consumption. In (Dou, H. et. al., 2022), an anonymous privacy-preserving data aggregation approach for fog-enhanced IoT is provided, ensuring that sensitive data is secured. In this strategy, the pseudonym method is employed to give both anonymity and authenticity. The Paillier method is also employed in data aggregation for data privacy protection. This technique is effective for establishing real-time communication with devices with low resources. However, it is inapplicable to the creation of smart grids. (Ramya Shree A. N., 2022) proposes a context-aware privacy-preserving method for IoT-based Smart Cities that use Software Defined Networking (Choudhary, S. et. al. 2022). This method's implementation on a paradigm known as software defined networking (SDN) enables Smart city IoT. Data packets passing over the network are checked for any potential breaches of privacy. When sensitive data is encountered, the network is monitored by an SDN controller, which separates the data in a 70/30 ratio. After that, the first half of the data is routed over a secure network channel, and the second part is sent through VPN.

IoT sensitive data is safeguarded from being hacked and disclosed in the privacy-preserving IoT architecture. To restrict sensitive data access, it employs the homomorphic encryption approach. The data aggregation to addends does not make sensitive data accessible to hackers/attackers in this case. Here, end-to-end privacy-preserving data access is offered, and its assessment is done in terms of query processing time efficiency.

With Cognitive IoT, useful ideas may be derived from IoT device data, and truth discovery methodologies can be used to verify the veracity of the acquired data. As a result, no breach of privacy is more important in the design of the truth finding. To deal with this, (Daniel, E., et. al. 2019) developed LPDA, a lightweight framework for truth finding that ensures privacy in fog-based IoT systems. To avoid any privacy breaches, it uses the Paillier cryptosystem in conjunction with the one-way hash chain approach. It assures truth discovery by preventing the insertion of misleading data, resulting in lower computational and communication overhead.

Techniques for Enhancing Noise

By changing secret qualities with noise added to the original data, the identification of a specific person is prohibited. The four categories to which all noise addition methods belong are as follows:

- Laplace noise addition to database query results is a differential privacy approach.
- Data sampling technique: a new table is released that only contains sample data on the whole population.
- Adding/multiplying the value of a sensitive property to a randomized integer is a random-noise approach.
- Modification of a data subset with the introduction of ambiguity in the real data value is a data swapping approach.
- On the Internet of Things, a Reliable Third Party

A trajectory privacy-preserving technique for mobile IoT devices is given, which depends on a trusted anonymous server. The privacy of the user's location is safeguarded here.

For group users snapshot queries, fulfilment of the spatial k-anonymity criterion is required. Individual users' location privacy is safeguarded by the LBS provider's ability to withstand inference assaults. For continuing inquiries, a circular safe area construction concept is also developed. By employing the optimum average closest neighbour approach, the distance between users is maintained but the true location information of the users is hidden. (Sethi, K. et. al., 2021) describes an outsourced multi authority access control technique that uses attribute-based encryption. The Cipher text-policy attribute-based encryption approach is used to create a privacy-preserving algorithm that makes all attributes anonymous while still ensuring safe authentication. Computational burden is reduced when decryption computation is outsourced.

- (Quane, K., 2021) highlighted TTP engagement in this sector in order to reduce query response time in a road network as well as anticipate traffic by users.
- (Al-Dhubhani, R. et. al., 2018) presents a discussion on location anonymization in continuous LBS.
- (Yousefpoor, E., 2021) explains how to reduce clacking area in a particular method.
- (Xu, J., 2021) describes a central party that provides adoptive algorithms for obscuring user locations.
- K-anonymity approaches seem to be incorrect in a few cases of spatial user distributions (Belsis, P. et. al., 2014), and those methods that employ k-anonymity have numerous drawbacks (Liu, Y. et. al. 2019). As a result, (Domingo-Ferrer, J. et. al., 2005) describes a way for a user to determine the minimal k- anonymity level using TTP (Alblooshi, M. et. al., 2018).
- (Huberty, M. 2015) explains the systematic noise generation reliance, the merging process with the user's position prior to cloaking area construction, and finally the transformation of area coordinates (query location) to the LBS server.

Researchers that rely on k-anonymity use "Semantic Cloaking" (Sen, A. A. A., et. al. 2021) or "Movement Vector" (Angayarkanni, S. A., et. al. 2021) approaches. It considers user mobility in order to reduce area size and cost, as well as discussing the field where the k-anonymity approach is useless, as well as when cloaking area users, their nature, and kinds are not taken into account. This may be solved by creating a unique profile for each person.

Pros:

- SP protects the privacy of the user's identity and location.

Cons:

- Shifting trust from SP to TP is a problem.
- With the Cloaking approach, there is a TTP overhead.

Unsolved Issues:

- Users with no trust deal with third parties (TP).
- Creating a cloak area with no TTP by collaborating users.

Cache method with TTP (Fig 3)

Working with Information

The following procedures (Gao, Y. L. et. al. 2021) are used to secure personal/sensitive data in this case:

- encryption
- steganography
- a disturbance (noise addition) (Jayasingh, R., et. al., 2022)
- data that is constantly moving
- dispersion of data
- data deletion on a regular basis
- removing a personal component
- prevention of storage
- Minimization of data using data mining (Samanta, D. et. al., 2021) and statistical approaches (Bhattacharya, A. et. al., 2021).

When the service provider does not need detailed information, the above-mentioned key approaches are used to safeguard consumers from external threats. Because of this capability, it may be employed as a superior solution in a variety of applications, for instance, in an energy system, determining the power usage over a certain time period.

Pros:

Figure 3. Cache Approach with TTP

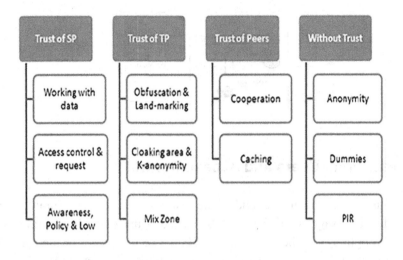

- As reduced/modified data, a few ways are employed to offer privacy from all attackers to some level.
- Some technologies, such as encryption and steganography (Pramanik, S. et. al., 2014), rely only on privacy-preservation against external assaults due to the need of a trusted partner.

Cons:

- The major danger here is that incomplete privacy protection causes an incorrect result, while some other approaches need complete faith in the SP.
- Additionally, applying encryption, steganography, or dispersed data has a negative impact on performance.

Unsolved Issues:

- The authors are looking for a suitable way for detecting and evaluating personal information.
- Encryption and steganography researchers are looking for novel techniques with great performance and privacy.

Figure 4. Technique of obfuscation

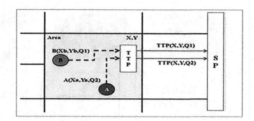

Requests From Users and Access Control

It allows users to read, change, and delete material that has been submitted to the server. For SP's usage, some data may be locked or unlocked. In addition, an alert (notice and option) may be sent, and the SP can be requested for the necessity of privilege for own data access, as well as authorisation for repetitive requests. For decreasing connections to SP, a connection disables and enable function is also available.

Pros:

- Easier
- Protects users' privacy from outside threats.

Cons:

- Here, SP's confidence is expected.
- It's still an open question.
- SP's assurance in enforcing the announced policy

Policies, Public Awareness, and Legislation

Different users must be aware of important information about privacy protection rights. Furthermore, various companies must adhere to current rules and policies.

Pros:

- Finding all of these is necessary in order to protect users' privacy from all types of attacks.

Cons:

- All of them are hypothetical; strong access authorization is required to verify the accuracy of the company's implementation of these regulations.

Unresolved problems:

- Developing new methods for keeping track on SP's behavior.
- Searching for a program that can identify the number of privacy violations in any software or code.

Various methods may need strong government backing in order for service providers to be properly monitored and supervised. As an example, some Android (Purkayastha, K.D., 2021) apps need permission to access the camera and microphone, yet this might result in video/user voice recordings being abused with no warning or indication.

Obfuscation and Land-Marking

The obfuscation process is shown in Figure 4. Here, mathematical and transformation functions are employed to adjust sensitive information as for various locations, whereas in land-marking, known places are utilised as the query location instead of actual coordinates.

In (Ifzarne, S. et. al., 2021), a new measure based on obfuscation is added, allowing users to indicate their privacy preferences using a new style.

(Croft, W. L. et. al., 2021) proposes a methodology for achieving computational efficiency in the obfuscation process.

In (Wee, H. et. al. 2021), the new Casper technology (Bandara, E., 2021) is introduced, which conceals the user's identity by disguising its real location.

In (Luh, R. et. al. 2017), a novel approach for protecting location privacy utilising semantics-aware obfuscation strategies is described.

Obfuscation is a fantastic solution for single and sequential requests. However, it necessitates a greater attention on the efficiency of its algorithms, hence boosting result accuracy and taking into account attacker skills such as map-related information, sketching users' movement tracks, and detecting data noise.

Pros:

- The privacy of a user's location is really protected.
- Ability to change a few personal details such as age, salary, and so on.

Cons:

Figure 5. Zone mixing method

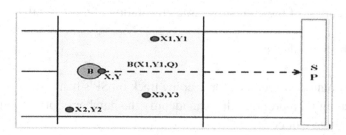

- There are a variety of effects on performance as well as reaction accuracy.

Zone of Mixing

As illustrated in Fig. 5, numbers of zones are generated by area partition, and users in each zone choose a new nickname.

- The limits of the mix-zone approach, which takes into consideration the road network and rectangular form, are explained in (Benarous, L. et. al. 2022). It offers a variety of methods for increasing the efficiency of an area by combining engineering and statistical behavior.
- Fine outputs are created with the least amount of computer complexity in (Nadeem, M. 2021) by processing a mathematical pattern (Sharma, S. K., 2021) that optimizes the mix-zone.
- (Ye, X. et. al. 2021) propose an innovative strategy to dealing with mix-zone by viewing real-world roads as heterogeneous in combination with traffic density.

Mix-zone is one of the most effective approaches for preventing user movement tracking. However, distinct areas to examine in order to increase its efficiency include traffic, not equal, noise addition in each zone's time, and traffic, not equal, pathways. Pros:

- In compared to other strategies that simply employ one nickname, this technique is more refined.
- Provides a higher level of secrecy from an attacking agent.

Cons:

- When a person uses the same internet connection to connect to a server, this is inappropriate.
- It implies that in order to be tracked, SP must rely on the user's genuine IP address (Bahashwan, A. A., et. al., 2021).

Privacy Information Retrieval (PIR)

Here, the user submits a query and receives chosen records as a response from the SP/DB, despite the fact that the SP is unaware of the identification record. It is accomplished by specifying a collection of rows rather than a single row. As a result, the user demands quick retrieval of the needed record from this collection. In certain cases, retrieval may be performed from several servers with multiple encrypted databases. This is seen in Fig. 6.

- (Wu, Z. et. al. 2021) highlighted many methods for protecting privacy from attackers who do not have TTP.
- Hardware-based (mid-server) PIR and obfuscation for load discharge were explored in (Alagic, G. et. al 2021) and (Bulat, R. et. al. 2022), which make them appropriate for real-time applications.

Pros:

- Provides optimum protection against adversaries and attackers.

Cons:

- Implementation is costly.
- Using such a protocol adds to the calculation and link overhead.
- The need for encryption and multi-server management, which seems to be unfeasible.

Unresolved Problems:

- Look for a better encryption algorithm and consider sharing or dividing the encryption key.

PIR is an excellent approach for retrieving user-required data without disclosing server specifications. It has an impact on system performance and necessitates the implementation of a few rules in the server provider, which is not always doable. As

Figure 6. Illustration of the PIR Technique (Encryption)

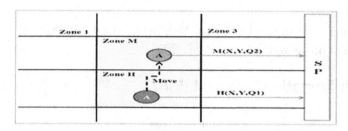

a result, this methodology is combined with TTP in order to improve the efficiency of these two methods.

Dummies

As shown in Figure 7, user 'A' sends a series of fraudulent inquiries, including a legitimate inquiry. From SP, this series of inquiries has various locations or kinds of queries to obfuscate the genuine query inside them.

Pros:

- SP as well as privacy protection from outside attacks.

Cons:

- Difficulty in consistently producing decent dummies.
- The attacker may be able to determine real searches by tracking the user query over a period of time.

Unresolved Problems:

Figure 7. Techniques for dummies

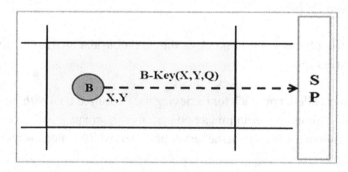

- Dummy creation that is very efficient.
- (Luo, J. N. et. al. 2019) covers the production of dummies before sending a query to the LBS server.
- (Siddiqui, S. et. al. 2021) examines the problem of dummy creation and offers tools to help users accomplish it.
- In (Van den Broeck, J. et. al. 2021), a technique for dummy site selection for dummy creation using probability of regions and entropy metrics is presented.

For discrete queries, this approach works well. However, work on clever dummy generation has to be improved in order for it to be used in future requests, and their ability to be disclosed by the server provider has been made too tough.

Collaboration Among Peers

As shown in Figure 8, direct collaboration among users/peers takes happen in a variety of methods for privacy preservation, including sharing query responses, exchanging them among users in a crowd, and shielding them from TTP/SP via cooperation.
Pros:

- It may be used to hide a user's identity from a different attacker/SP.
- As much as possible, reducing the number of communications with it

Cons:

- The necessity for all users to be in the same area is one of the method's drawbacks (wireless connection)
- It's a question of user trust.

Unresolved problems:

- User trustworthiness
- Peers' reputation

- (Abi Sen A. A. 2018) highlights user operational difficulties in P2P through POIs/sub-queries exchange.
- Prior to the SP's connection, (Guan J. et. al. 2021) integrates cryptography with cooperative mechanism through query exchange/cryptographic key exchange.

Figure 8. Illustration of a collaboration method (swapping)

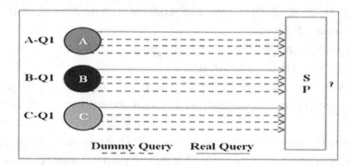

- The development of an obfuscation area by users creates great uniformity in (Gay, R. et. al. 2021).
- (Ma, Y. et al. 2021) is built on cooperation as well; however the swapping is done on the clocking areas.
- (Sen, A. A. A. et. al., 2021) presented a novel cooperation strategy based on assembling a group of peers with the same POIs to save overhead, hence reducing SP's number of connections and authenticating the replies.
- (Wernke, M. et. al. 2014) suggested masking users' location from SP by allowing them to collaborate without using TTP.

When there is a lack of confidence between the user and any other party, peer collaboration is a necessary strategy. As a result, other users collaborate for the purpose of benefit sharing and privacy enhancement. However, achieving the setup for peer-to-peer communication is not simple. There are several algorithms for collaboration that may be used and suggested. Furthermore, combining this procedure with other ways may lead to the development of new strategies.

Caching

As may be seen in Fig. 9, it employs a caching approach for saving certain query responses, which it then reuses for subsequent queries. It reduces the amount of connections available to SP, posing a danger. This strategy may be integrated with others, necessitating the creation of a unique infrastructure.

Pros:

- LBS's number of connections is reduced.
- Enhancement of user privacy

Figure 9. An illustration of a caching mechanism

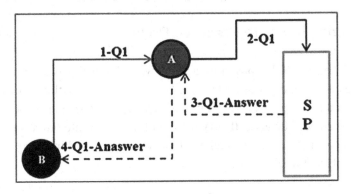

Cons:

- This strategy is possible in a smart city with smart infrastructure; nevertheless, it needs user faith in the cache region.
- In the missing situation, there is a link between the users and the SP.

Unresolved problems:

- Improves cache hit ratio, resulting in increased privacy.

Crowd information, which enables users to cache their query replies, will make it easier for other users to have the same inquiry in the future without having to connect to SP.

- A TTP-based cache-cloak system was presented in a trusted server/memory. It generates k-anonymity in real time and caches a small number of future query replies.
- Cache setup at each cell's access point is proposed, and it also enhances cache hit-ratio by using smart dummies during user and SP connection.

By reducing the number of interactions with server providers, the caching approach may be used more often to increase privacy and speed. However, it requires a specialised configuration, such as a smart city, or its integration with other methodologies. It's also utilised to identify alternative tools for indirect peer collaboration, which leads to a reduction in privacy strategies.

OTHER INTERNET OF THINGS TECHNIQUES

Using Game Theory to Preserve Privacy

(Kaushik, D. et. al. 2021) proposes a framework for IoT privacy protection that incorporates social interaction and is based on game theory. In this case, the analysis of complex interactions in a network involving a service provider, an adversary, and a user is performed. The game theory methodology looks into online users' privacy leaks in order to analyse their behaviour. This concept uses a third-party game model to assure confidential data trade.

Blockchain-based Privacy Protection

Mobile crowd sensing is another term for crowd sensing. It allows mobile users to collect, share, and compute data in order to seek incentives. The location privacy of mobile users is jeopardised in crowd sensing. (Bandyopadhyay, S. et. al. 2021) demonstrated a block chain-based location privacy-preserving crowd sensing technology. The anonymous nature of block chain technology is employed to safeguard user privacy. As a result, it makes it harder for attackers to acquire transaction data, ensuring that the system is safe against re-identification attacks.

Because of its decentralised nature and cryptographic technology, block chain has several benefits. Thin-client systems are not supported here; for example, IoT devices cannot be complete nodes in a block chain. As a result, maintaining user privacy in a block chain-based IoT system is a significant difficulty. This problem may be solved by implementing "PTAS: Privacy-preserving Thin-client Authentication Scheme" in PKI, which uses block chain as presented in (Pandey, P. et. al. 2021), allowing the thin-client system to function as a complete node with a private information retrieval (PIR) system. PIR is responsible for protecting the identification of nodes as well as maintaining system privacy. Furthermore, there are (m-1) private authentication systems, which increase system security, and privacy is assured even if (m-1) numbers of nodes collude at the same time.

Fog Computing for Privacy Protection

In a fog-assisted IoT system, (Pramanik, S. et. al. 2022) proposed a framework for privacy-preserving data search. Fog nodes collect data from IoT devices, which is then stored in a fog-based cloud system. In this approach, fog nodes are used to discover the info that users are looking for. There are two encryption algorithms that are utilised, both of which are searchable.

- Encryption that can be searched
- Searchable encryption with the help of a semi-trusted fog node.

The first technique attempts to reduce computing costs while also enabling IoT devices that are not connected to the internet and the second way provides fine-grain access search across fog nodes while also protecting user privacy via the use of multiple authentication credentials.

IoT devices generate massive amounts of data, which a cloud platform can efficiently handle. Even yet, there are various problems in the cloud environment, including power consumption, access time, privacy, data location, and resource efficiency. In (Pramanik, S. et. al. 2022), an IoT-oriented data placement approach is created to address the above-mentioned conflicts in the cloud environment. This paper proposes a privacy-aware data placement system that prevents privacy leakage by applying various privacy constraints at the host end.

The main issue in a cloud environment is securing data from IoT applications. The Chinese Remainder Theorem (CRT) is used to create a storage mechanism for securely storing user data. A CRT-based group key management technique was established for accessing cloud-encrypted data. For data privacy protection, a variety of encryption and decryption stages are utilised. However, with key sizes less than or equal to 512 bits, it is inefficient.

Encryption Based on Chaos for Privacy Preservation

E-Healthcare is a prominent application area for IoT. Data on a patient's health and medical history is often sensitive. In E-Healthcare, the privacy of the patient is paramount. (Kaur, G. et. al. 2022) proposed a chaos-based encryption cryptosystem to protect patient privacy. Medical images are encrypted using a fast probabilistic image encryption (Gupta, A.et. al. 2022) cryptosystem. Medical key-frames are extracted from wireless capsule endoscopy and kept safe here as well. Confusion and diffusion techniques are employed to carry out symmetric block encryption in this method.

In contrast to privacy-preserving IoT systems, which need a weaker identity, IoT systems with high security features demand a stronger identity. (Zhang, Y. et. al. 2021) describes a privacy-preserving authentication technique for IoT systems that achieves a good mix of privacy and security. It was created to accommodate IoT end devices with weaker identities. For secure communication among IoT devices, a secret sharing technique is utilised. For the development of the authentication protocol, it also employs a short group signature technique.

Figure 10. IoT architecture in three layers

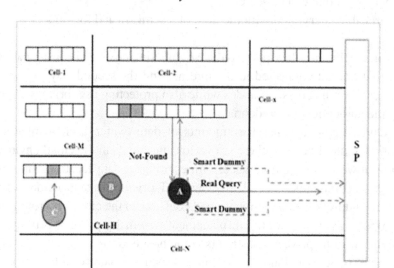

IOT ENVIRONMENTS THAT PROTECT YOUR PRIVACY

The application of current Internet standards on various smart devices can simplify the projected situation from an IoT perspective. However, maintaining privacy in IoT applications requires changes/expansions to normal internet protocol security methods. This section will look at how to maintain privacy over a three-layered IoT infrastructure. Look at Figure 10.

Privacy-Preservation at the Device Layer

This device layer contains many physical resources for collecting/controlling data, such as sensors and actuators. However, they are resource-constrained and very diverse, making privacy-preserving techniques challenging to implement. In (Kuzlu, M. et. al. 2021), many attacks on IoT devices are covered, including side channel attack (SCA), fake node, DoS attack, mass node authentication issue, timing attack, node capture, malicious data, routing threats, and replay attack. As a result, the following security techniques should be considered while designing the device layer:

- Data integrity, secrecy, privacy protection, and authenticity are all provided by cryptography technology. Secure communication methods using hash values and digital signatures are used to ensure data integrity.

- Authentication and access control: It protects users' privacy from a variety of open and unauthorized access points. Selective RFID jamming is used as a low-cost tag access control system in (Munhoz, M. et. al. 2021).
- Data encryption: It protects data exchange and guarantees that it is sent safely. (Haider, M. I. 2021) proposes a nonlinear key technique based on displacement calculations which encrypt information. This algorithm requires little processing power in order to achieve higher security and a faster data transmission rate.
- IPsec protocol provides both encryption and authentication in a secure channel. (Dushyamt, K. et. al. 2022) proposes a 6LoWPAN/IPsec extension to provide security to IoT devices. In this paper, it is shown that IPSec outperforms conventional IEEE802.15.4 link-layer security in an IoT scenario.

Privacy-Preservation at the Platform/Infrastructure Layer

It is similar to the network layer of the OSI model. This is where intelligent data pre-processing is integrated in order to reduce resource needs at the Application Layer. Accessing network without authorisation, Man-in-the-Middle attacks, eavesdropping, DoS attacks, confidentiality and integrity damage are all examples of frequent security challenges with integrity and confidentiality. Different network protocols that existed before are used to provide high-security measures; but, in a resource-constrained context with M2M communication, they are ineffective. As a consequence, present security mechanisms on various IoT devices either result in insufficient security or inapplicability, resulting in a non offence rather than links between machines. As a result, this network diversity compromises security, interoperability, and network coordination, posing a security risk (Pradhan, D. et. al. 2022). In order to completely integrate in the IoT ecosystem, emerging IoT-based security solutions should take into consideration the following security requirements:

- Establishing security routing, PKI (Public Key Infrastructure), end-to-end authentication, intrusion detection systems, key agreement mechanisms, and WPKI for wireless, among other things.
- Using network virtualization to simplify network administration and eliminate any potentially ineffective operations.
- Adoption of IPv6 as a standard network layer protocol to enable security techniques evolved from it.

Privacy-Preservation at the Application Layer

It is made up of two layers:

- Analytical services and edge computing are run by the support layer.
- The application service layer is in charge of providing critical IoT infrastructure support.

Based upon different requirements, IoT applications are of very adaptable and diverse, so providing standard support is much challenging too. Different applications focus on different specific area having the need of unique data collection that may need a number of security measures. Hence, application layer's security considerations/requirements are different from other 2 stack layers in following aspect:

Technical

Anonymous authentication, fingerprint technology, homomorphic and threshold cryptography, and digital watermarking are all examples of cryptography.

Certification transfer technology, symmetric and asymmetric cryptography are among the key agreements.

Non-technical

- Users' understanding of private data collecting, possible threats, IoT service safety, and how to prevent private information breaches.
- Security management includes resource protection, password management, and physical security data, among other things.

Challenges Arising with the Implementation of Security and Privacy Concerns in Data Aggregation Method in IoT

Physical things are usually idled or discarded even during post-working period. Most of the time, their owners disregard their privacy and security issues, leaving them open to attack targets (like being seized, hacked, or recycled by attackers) and facing major security dangers and privacy leakage hazards. Meanwhile, advances in microelectronics, biometric information technologies, and cryptography provide promising prospects for protection and privacy-preserving solutions. The linked studies are summarised in trying to sort out the privacy and security difficulties encountered by physical items and to explore future prospects.

Challenges

There are six dimensions to the privacy and security problems that physical things may encounter in the omnipresent IoT.

- Edge computing is a kind of computing that is used at the edge of the network. IoT has a slew of new privacy and security flaws. When the link between the cloud and the perception network gets closer, the edge node is in proximity to the data source and the local data centres, making it a target of attack and exposing confidential data.

- The IoT-based ubiquitous computing framework, that is entirely disseminated and immensely dynamic, must strike stability between privacy, service and security. Superior service and security constantly need accurate private data, although superior security needs lesser data exposure, making current static privacy and security explanations ineligible of having the dynamic privacy requirements of various objects.

- Because IoT-dependent ubiquitous computing typically necessitates data exchange and data usage across privacy domains or units of Internet of Things, sensitive data allocation and usage is fraught with security and privacy concerns. In what way may a data owner, for example, efficiently outsource the data while maintaining privacy?

- A physical object is frequently required to connect to an unknown thing in a dissimilar privacy area. The cornerstone of their privacy and security is the formation of a fundamental trust connection between two physical things which do not recognize each other.

- The foundation for guaranteeing the privacy and security of physical things and their owners is to prohibit unauthorised people from accessing them. Although utilising biometric data to link physical things to their owners is a popular option, there are still a slew of additional security and privacy issues to be addressed.

- As blockchain technology advances, there appears to be a tendency toward using blockchain to govern physical assets. Despite this, the public ledger forces users to confront a data security and privacy threat.

CONCLUSION

As the privacy problems in IoT devices are the most susceptible of all the challenges, such a rapid increase in IoT use has an influence on the privacy and device life spans. All the devices interacted one to one, resulting in a heavy traffic that might reduce the life of unit nodes. Furthermore, transmitting data on a regular basis raises the chances of an attacker assaulting the system. The security problems may be exacerbated by such traffic. The use of an aggregator in the scheme as a link between end nodes and the sink may be able to solve these issues. The authors may use an intermediate device to aggregate all of the individual sensor data and provide it to the sink in a single packet in any system with many sensors or nodes and a central controller or sink. The term given to such a device or component is an aggregator. Data aggregation is used to reduce the traffic and communication costs. In general, this method aids in the extension of a node's life while simultaneously lowering the network transmission. The researchers don't have to worry about privacy when we employ an aggregator between the node and the sink since only the genuine user knows how the data is obtained in the aggregator. As a consequence, robust aggregation and system security are required, among other things, to reduce communication overhead, prolong the life of the node, and enable secure and easy communication.

REFERENCES

Abi Sen, A. A., Alawfi, I. M. M., Aloufi, H. F., Bahbouh, N. M., & Alsaawy, Y. (2021). Comparison among Cooperation, Anonymity and Cloak Area approaches for Preserving Privacy of IoT. *2021* 8th International Conference on Computing for Sustainable Global Development *(INDIACom)*, 413-416. doi: 10.1109/INDIACom51348.2021.00073

Abi Sen, A. A., Eassa, F. A., & Jambi, K. (2018). Preserving privacy in internet of things: A survey. *Int. J. Inf. Tecnol.*, *10*, 189–200. doi:10.1007/s41870-018-0113-4

Al-Dhubhani, R., & Cazalas, J. M. (2018). An adaptive geo-indistinguishability mechanism for continuous LBS queries. *Wireless Networks*, *24*(8), 3221–3239. doi:10.100711276-017-1534-x

Alagic, G., Brakerski, Z., Dulek, Y., & Schaffner, C. (2021). Impossibility of Quantum Virtual Black-Box Obfuscation of Classical Circuits. In T. Malkin & C. Peikert (Eds.), Lecture Notes in Computer Science: Vol. 12825. Advances in Cryptology – CRYPTO 2021. CRYPTO 2021. Springer. https://doi.org/10.1007/978-3-030-84242-0_18.

Alblooshi, M., Salah, K., & Alhammadi, Y. (2018). Blockchain-based Ownership Management for Medical IoT (MIoT) Devices. *2018 International Conference on Innovations in Information Technology (IIT)*, 151-156. 10.1109/INNOVATIONS.2018.8606032

Angayarkanni, S. A., Sivakumar, R., & Ramana Rao, Y. V. (2021). Hybrid Grey Wolf: Bald Eagle search optimized support vector regression for traffic flow forecasting. *Journal of Ambient Intelligence and Humanized Computing*, *12*(1), 1293–1304. doi:10.100712652-020-02182-w

Bahashwan, A. A., Anbar, M., Abdullah, N., Al-Hadhrami, T., & Hanshi, S. M. (2021). Review on Common IoT Communication Technologies for Both Long-Range Network (LPWAN) and Short-Range Network. In F. Saeed, T. Al-Hadhrami, F. Mohammed, & E. Mohammed (Eds.), *Advances on Smart and Soft Computing. Advances in Intelligent Systems and Computing* (Vol. 1188). Springer. doi:10.1007/978-981-15-6048-4_30

Bandara, E., Shetty, S., & Mukkamala, R. (2021). *Casper: a blockchain-based system for efficient and secure customer credential verification. J Bank Financ Technol.* doi:10.100742786-021-00036-3

Bandyopadhyay, S., Goyal, V., Dutta, S., Pramanik, S., & Sherazi, H. H. R. (2021). Unseen to Seen by Digital Steganography. In S. Pramanik, M. M. Ghonge, R. Ravi, & K. Cengiz (Eds.), *Multidisciplinary Approach to Modern Digital Steganography*. IGI Global.

Baouya, A., Chehida, S., Cantero, M., Millet, M., Bensalem, S., & Bozga, M. (2021). Formal Modeling and Simulation of Collaborative Intelligent Robots. In *Advances in Service-Oriented and Cloud Computing. ESOCC 2020. Communications in Computer and Information Science* (Vol. 1360). Springer. doi:10.1007/978-3-030-71906-7_4

Belsis, P., & Pantziou, G. (2014). A *k*-anonymity privacy-preserving approach in wireless medical monitoring environments. *Personal and Ubiquitous Computing*, *18*(1), 61–74. doi:10.100700779-012-0618-y

Benarous, L., & Kadri, B. (2022). Obfuscation-based location privacy-preserving scheme in cloud-enabled internet of vehicles. *Peer-to-Peer Networking and Applications*, *15*(1), 461–472. doi:10.100712083-021-01233-z

Bhattacharya, A., Ghosal, A., Obaid, A. J., Krit, S., Shukla, V. K., Mandal, K., & Pramanik, S. (2021). Unsupervised Summarization Approach with Computational Statistics of Microblog Data. Methodologies and Applications of Computational Statistics for Machine Intelligence, 23-37.

Bułat, R., & Ogiela, M. R. (2022). Personalized Cryptographic Protocols - Obfuscation Technique Based on the Qualities of the Individual. In L. Barolli, H. C. Chen, & T. Enokido (Eds.), Advances in Networked-Based Information Systems. NBiS 2021. Lecture Notes in Networks and Systems (Vol. 313). Springer. https://doi.org/10.1007/978-3-030-84913-9_19.

Choudhary, S., Narayan, V., Faiz, M., & Pramanik, S. (2022). Fuzzy Approach-Based Stable Energy-Efficient AODV Routing Protocol in Mobile Ad hoc Networks. In *Software Defined Networking for Ad Hoc Networks* (pp. 125–139). EAI/Springer Innovations in Communication and Computing. doi:10.1007/978-3-030-91149-2_6

Croft, W. L., Sack, J. R., & Shi, W. (2021). Obfuscation of images via differential privacy: From facial images to general images. *Peer-to-Peer Networking and Applications*, *14*(3), 1705–1733. doi:10.100712083-021-01091-9

Daniel, E., & Vasanthi, N. A. (2019). LDAP: A lightweight de-duplication and auditing protocol for secure data storage in cloud environment. *Cluster Computing*, *22*(S1), 1247–1258. doi:10.100710586-017-1382-6

Domingo-Ferrer, J., & Torra, V. (2005). Ordinal, Continuous and Heterogeneous *k*-Anonymity Through Microaggregation. *Data Mining and Knowledge Discovery*, *11*(2), 195–212. doi:10.100710618-005-0007-5

Dou, H., Chen, Y., Yang, Y., & Long, Y. (2022). A secure and efficient privacy-preserving data aggregation algorithm. *Journal of Ambient Intelligence and Humanized Computing*, *13*(3), 1495–1503. doi:10.100712652-020-02801-6

Dushyant, K., Muskan, G., Gupta, A., & Pramanik, S. (2022). Utilizing Machine Learning Deep Learning in Cyber security: An Innovative Approach. In M. M. Ghonge, S. Pramanik, R. Mangrulkar, & D. N. Le (Eds.), *Cyber security and Digital Forensics*. Wiley.

Dutta, S., Pramanik, S., & Bandyopadhyay, S. K. (2021). Prediction of Weight Gain during COVID-19 for Avoiding Complication in Health. *International Journal of Medical Science and Current Research*, *4*(3), 1042–1052.

Elbasani, E., Siriporn, P., & Choi, J. S. (2020). A Survey on RFID in Industry 4.0. In G. Kanagachidambaresan, R. Anand, E. Balasubramanian, & V. Mahima (Eds.), *Internet of Things for Industry 4.0. EAI/Springer Innovations in Communication and Computing*. Springer. doi:10.1007/978-3-030-32530-5_1

Gao, Y. L., Chen, X. B., Xu, G., Liu, W., Dong, M.-X., & Liu, X. (2021). A new blockchain-based personal privacy protection scheme. *Multimedia Tools and Applications*, *80*(20), 30677–30690. doi:10.100711042-020-09867-6

Gay, R., Jain, A., Lin, H., & Sahai, A. (2021). Indistinguishability Obfuscation from Simple-to-State Hard Problems: New Assumptions, New Techniques, and Simplification. In A. Canteaut & F. X. Standaert (Eds.), Lecture Notes in Computer Science: Vol. 12698. Advances in Cryptology – EUROCRYPT 2021. EUROCRYPT 2021. Springer. https://doi.org/10.1007/978-3-030-77883-5_4.

Guan, J., & Zhandry, M. (2021). Disappearing Cryptography in the Bounded Storage Model. In K. Nissim & B. Waters (Eds.), Lecture Notes in Computer Science: Vol. 13043. Theory of Cryptography. TCC 2021. Springer. https://doi.org/10.1007/978-3-030-90453-1_13.

Gupta, A., Verma, A., & Pramanik, S. (2022). Security Aspects in Advanced Image Processing Techniques for COVID-19. In S. Pramanik, A. Sharma, S. Bhatia, & D. N. Le (Eds.), *An Interdisciplinary Approach to Modern Network Security*. CRC Press.

Haider, M. I., Ali, A., & Shah, D. (2021). Block cipher's nonlinear component design by elliptic curves: an image encryption application. *Multimed Tools Appl*, *80*, 4693–4718. doi:10.1007/s11042-020-09892-5

Huberty, M. (2015). Awaiting the Second Big Data Revolution: From Digital Noise to Value Creation. *Journal of Industry, Competition and Trade*, *15*(1), 35–47. doi:10.100710842-014-0190-4

Ifzarne, S., Hafidi, I., & Idrissi, N. (2021). Secure Data Collection for Wireless Sensor Network. In M. Ben Ahmed, S. Mellouli, L. Braganca, B. Anouar Abdelhakim, & K. A. Bernadetta (Eds.), *Emerging Trends in ICT for Sustainable Development. Advances in Science, Technology & Innovation (IEREK Interdisciplinary Series for Sustainable Development)*. Springer. doi:10.1007/978-3-030-53440-0_26

Jayasingh, R. (2022). Speckle noise removal by SORAMA segmentation in Digital Image Processing to facilitate precise robotic surgery. *International Journal of Reliable and Quality E-Healthcare*, *11*(1), 1–19. Advance online publication. doi:10.4018/IJRQEH.295083

Kaur, G., Agarwal, R., & Patidar, V. (2022). Color image encryption scheme based on fractional Hartley transform and chaotic substitution–permutation. *The Visual Computer*, *38*, 1027–1050. https://doi.org/10.1007/s00371-021-02066-w

Kaur, M. J., Mishra, V. P., & Maheshwari, P. (2020). The Convergence of Digital Twin, IoT, and Machine Learning: Transforming Data into Action. In M. Farsi, A. Daneshkhah, A. Hosseinian-Far, & H. Jahankhani (Eds.), *Digital Twin Technologies and Smart Cities. Internet of Things (Technology, Communications and Computing)*. Springer. doi:10.1007/978-3-030-18732-3_1

K.aushik, D., Garg, M., Gupta A., & Pramanik, S., (2021). Application of Machine Learning and Deep Learning in Cyber security: An Innovative Approach. In *Cybersecurity and Digital Forensics: Challenges and Future Trends*. Wiley.

Kuzlu, M., Fair, C., & Guler, O. (2021). Role of Artificial Intelligence in the Internet of Things (IoT) cybersecurity. *Discov Internet Things*, *1*, 7. doi:10.1007/s43926-020-00001-4

Liu, Y., & Zhao, Q. (2019). E-voting scheme using secret sharing and K-anonymity. *World Wide Web (Bussum)*, *22*(4), 1657–1667. doi:10.100711280-018-0575-0

Luh, R., Marschalek, S., Kaiser, M., Janicke, H., & Schrittwieser, S. (2017). Semantics-aware detection of targeted attacks: A survey. *J Comput Virol Hack Tech*, *13*(1), 47–85. doi:10.100711416-016-0273-3

Luo, J. N., Yang, M. H., & Tsai, K. Y. (2019). A geographic map-based middleware framework to obfuscate smart vehicles' locations. *Multimed Tools Appl*, *78*, 28877–28902. doi:10.1007/s11042-019-7350-9

Ma, Y., Li, Y., Zhang, Z., Zhang, R., Liu, L., & Zhang, X. (2021). A Classic Multi-method Collaborative Obfuscation Strategy. In Y. Tan, Y. Shi, A. Zomaya, H. Yan, & J. Cai (Eds.), Data Mining and Big Data. DMBD 2021. Communications in Computer and Information Science (Vol. 1454). Springer. https://doi.org/10.1007/978-981-16-7502-7_10.

Mandal, A., Dutta, S., & Pramanik, S. (2021). Machine Intelligence of Pi from Geometrical Figures with variable Parameters usingSCILab. In *Methodologies and Applications of Computational Statistics for Machine Intelligence*. IGI Global. doi:10.4018/978-1-7998-7701-1.ch003

Munhoz, A. M., Chala, L., & Melo, G. d. (2021). Clinical and MRI Evaluation of Silicone Gel Implants with RFID-M Traceability System: A Prospective Controlled Cohort Study Related to Safety and Image Quality in MRI Follow-Up. *Aesthetic Plastic Surgery*, *45*, 2645–2655. https://doi.org/10.1007/s00266-021-02355-8

Nadeem, M. (2021). Board Gender Diversity and Managerial Obfuscation: Evidence from the Readability of Narrative Disclosure in 10-K Reports. *Journal of Business Ethics*. Advance online publication. doi:10.100710551-021-04830-3

Nakai, T., Misawa, Y., & Tokushige, Y. (2021). How to Solve Millionaires' Problem with Two Kinds of Cards. *New Generation Computing*, *39*, 73–96. https://doi.org/10.1007/s00354-020-00118-8-m

Pandey, B. K., Pandey, D., Wairya, S., Agarwal, G., Dadeech, P., Dogiwal, S. R., & Pramanik, S. (2022). Application of Integrated Steganography and Image Compressing Techniques for Confidential Information Transmission. In Cyber Security and Network Security. Wiley. doi:10.1002/9781119812555.ch8

Pandey, P., & Litoriya, R. (2021). Securing E-health Networks from Counterfeit Medicine Penetration Using Blockchain. *Wireless Personal Communications*, *117*, 7–25. https://doi.org/10.1007/s11277-020-07041-7

Pradhan, D., Sahu, P. K., Goje, N. S., Ghonge, M. M., & Pramanik, S. (2022). Security, Privacy, Risk, and Safety Toward 5G Green Network (5G-GN). In Cyber Security and Network Security. Wiley. doi:10.1002/9781119812555.ch9

Pradhan, D., Sahu, P. K., Goje, N. S., Myo, H., Ghonge, M. M., Tun, M., Rajeswari, R., & Pramanik, S. (2022). Security, Privacy, Risk, and Safety Toward 5G Green Network (5G-GN). In Cyber Security and Network Security. Wiley. doi:10.1002/9781119812555.ch9

Pramanik, S., & Bandyopadhyay, S. K. (2013). Application of Steganography in Symmetric Key Cryptography with Genetic Algorithm. *International Journals of Engineering and Technology, 10*, 1791–1799.

Pramanik, S., & Bandyopadhyay, S. K. (2014). Image Steganography Using Wavelet Transform and Genetic Algorithm. *International Journal of Innovative Research in Advanced Engineering, 1*, 1–4.

Pramanik, S., & Bandyopadhyay, S. K. (2014). *An Innovative Approach in Steganography*. Scholars Journal of Engineering and Technology.

Pramanik, S., Joardar, S., Jena, O. P., & Obaid, A. J. (2022). An Analysis of the Operations and Confrontations of Using Green IT in Sustainable Farming. *Al-Kadhum 2nd International Conference on Modern Applications of Information and Communication Technology*.

Pramanik, S., Pandey, D., Joardar, S., Niranjanamurthy, M., Pandey, B. K., & Kaur, J. (2022). An Overview of IoT Privacy and Security in Smart Cities. *ICCAPE 2022*.

Pramanik, S., & Raja, S. (2019). Analytical Study on Security Issues in Steganography. *Think-India, 22*(35), 106-114.

Pramanik, S., & Raja, S. (2020). A Secured Image Steganography using Genetic Algorithm. *Advances in Mathematics: Scientific Journal, 9*(7), 4533–4541.

Pramanik, S., Sagayam, K. M., & Jena, O. P. (2021). Machine Learning Frameworks in Cancer Detection. ICCSRE 2021.

Purkayastha, K. D., Mishra, R. K., Shil, A., & Pradhan, S. N. (2021). IoT Based Design of Air Quality Monitoring System Web Server for Android Platform. *Wireless Personal Communications, 118*(4), 2921–2940. doi:10.100711277-021-08162-3

Sahoo, K. C., & Pati, U. C. (2020). IoT based intrusion detection system using PIR sensor. *2017 2nd IEEE International Conference on Recent Trends in Electronics, Information & Communication Technology (RTEICT)*, 1641-1645. 10.1109/RTEICT.2017.8256877

Samanta, D., Dutta, S., Galety, M. G., & Pramanik, S. (2021). A Novel Approach for Web Mining Taxonomy for High-Performance Computing. *The 4th International Conference of Computer Science and Renewable Energies (ICCSRE'2021)*. 10.1051/e3sconf/202129701073

Sarkar, M., Pramanik, S., & Sahnwaj, S. (2020). Image Steganography Using DES-DCT Technique. *Turkish Journal of Computer and Mathematics Education*, *11*(2), 906–913.

Sen, A. A. A., & Yamin, M. (2021). Advantages of using fog in IoT applications. *Int. J. Inf. Tecnol.*, *13*, 829–837. doi:10.1007/s41870-020-00514-9

Shankar, K., Elhoseny, M., Kumar, R. S., Lakshmanaprabu, S. K., & Yuan, X. (2020). Secret image sharing scheme with encrypted shadow images using optimal homomorphic encryption technique. *Journal of Ambient Intelligence and Humanized Computing*, *11*(5), 1821–1833. doi:10.100712652-018-1161-0

Sharma, S. K., & Ahmed, S. S. (2021). IoT-based analysis for controlling & spreading prediction of COVID-19 in Saudi Arabia. *Soft Computing*, *25*(18), 12551–12563. doi:10.100700500-021-06024-5 PMID:34305445

Siddiqie, S., Mondal, A., & Reddy, P. K. (2021). An Improved Dummy Generation Approach for Enhancing User Location Privacy. In Lecture Notes in Computer Science: Vol. 12683. Database Systems for Advanced Applications. DASFAA 2021. Springer. https://doi.org/10.1007/978-3-030-73200-4_33.

Siddiqui, I. F., Qureshi, N. M. F., Chowdhry, B. S., & Uqaili, M. A. (2020). Pseudo-Cache-Based IoT Small Files Management Framework in HDFS Cluster. *Wireless Personal Communications*, *113*(3), 1495–1522. doi:10.100711277-020-07312-3

Sinha, M., Chacko, E., Makhija, P., & Pramanik, S. (2021). Energy-Efficient Smart Cities with Green Internet of Things. In C. Chakraborty (Ed.), *Green Technological Innovation for Sustainable Smart Societies*. Springer. doi:10.1007/978-3-030-73295-0_16

Sinha, M., Chacko, E., Makhija, P., & Pramanik, S. (2021). Energy Efficient Smart Cities with Green IoT. In *Green Technological Innovation for Sustainable Smart Societies: Post Pandemic Era, C. Chakrabarty*. Springer.

Talaat, F. M., Saraya, M. S., Saleh, A. I., Ali, H. A., & Ali, S. H. (2020). A load balancing and optimization strategy (LBOS) using reinforcement learning in fog computing environment. *Journal of Ambient Intelligence and Humanized Computing*, *11*(11), 4951–4966. doi:10.100712652-020-01768-8

Van den Broeck, J., Coppens, B., & De Sutter, B. (2021). Obfuscated integration of software protections. *Int. J. Inf. Secur., 20*, 73–101. doi:10.1007/s10207-020-00494-8

Wee, H., & Wichs, D. (2021). Candidate Obfuscation via Oblivious LWE Sampling. In A. Canteaut & F. X. Standaert (Eds.), Lecture Notes in Computer Science: Vol. 12698. *Advances in Cryptology – EUROCRYPT 2021. EUROCRYPT 2021.* Springer. doi:10.1007/978-3-030-77883-5_5

Wernke, M., Skvortsov, P., & Dürr, F. (2014). A classification of location privacy attacks and approaches. *Pers Ubiquit Comput, 18*, 163–175. doi:10.1007/s00779-012-0633-z

Wu, Z., Li, G., Shen, S., Lian, X., Chen, E., & Xu, G. (2021). Constructing dummy query sequences to protect location privacy and query privacy in location-based services. *World Wide Web (Bussum), 24*(1), 25–49. doi:10.100711280-020-00830-x

Xu, J., Glicksberg, B. S., Su, C., Walker, P., Bian, J., & Wang, F. (2021). Federated Learning for Healthcare Informatics. *Journal of Healthcare Informatics Research, 5*(1), 1–19. doi:10.100741666-020-00082-4 PMID:33204939

Ye, X., Zhou, J., Li, Y., Cao, M., Chen, D., & Qin, Z. (2021). A location privacy protection scheme for convoy driving in autonomous driving era. *Peer-to-Peer Networking and Applications, 14*(3), 1388–1400. doi:10.100712083-020-01034-w

Yousefpoor, E., Barati, H., & Barati, A. (2021). A hierarchical secure data aggregation method using the dragonfly algorithm in wireless sensor networks. *Peer-to-Peer Networking and Applications, 14*(4), 1917–1942. doi:10.100712083-021-01116-3

Zhang, Y., Zou, J., & Guo, R. (2021). Efficient privacy-preserving authentication for V2G networks. *Peer-to-Peer Networking and Applications, 14*, 1366–1378. https://doi.org/10.1007/s12083-020-01018-w

Chapter 9
The Internet of Things:
Legal Realities

Theophilus Aigbogun
Marcus-Okoko & Co., Nigeria

ABSTRACT

The technology space has seen the emergence of several buzzwords, including but not limited to artificial intelligence, big data, the internet of things, and robotics. This chapter seeks to discuss the concept of the internet of things and break the very gigantesque details into understandable bits. It examines the concept of the internet of things from the standpoint of a legal practitioner who is practicing in any given jurisdiction and talks about the legal issues that attend the internet of things. Due to the fact that the internet of things refer to the interconnectivity of devices, the legal issue of data privacy and protection is discussed. Other issues like antitrust and 'who-bears-liability' are also equally discussed. A brief insight is equally given along the lines of how these legal issues preclude the smooth rolling out of these technologies in cross-border terms and how industry players are attempting to deal with the issues.

INTRODUCTION

Recently, the realms of Technology and Artificial intelligence has seen a quantum leap in its development. A major aspect of Technology that is presently tangibly operational in the world is the Internet of Things, and is not without its attendant legal issues. The Internet of Things is the interconnectivity of devices by way of the Internet, and an environment of data-collecting sensors with unique identifiers which can interact with each other through the transference of data. Oracle (2022).

DOI: 10.4018/978-1-7998-9312-7.ch009

The functionality of IoT devices has displaced the need for human-to-human or human-to-computer interaction because IoT devices can in, and by themselves, interpret data transferred between each other.

According to Statista, there are expected to be more than 30.9 billion IoT devices worldwide by 2025. Vailshery (2021). Also, about 127 IoT devices have been estimated to connect to the internet every second. Steward (2022) The advent of the Internet of Things has brought about several leaps and bounds. Devices like fitness bands, smart electric grids with sustainable energy solutions are all part of the IoT ecosystem. Diverse devices have become internet enabled, including, but not limited to smart refrigerators, smart lighting systems, and smart homes. A focal point to note is that these internet-enabled devices facilitate communication and interaction between themselves. These kinds of interactions are known as machine-to-machine or M2M interactions. These kinds of interactions substantially, or totally, eliminate the need for human intervention.

The Internet of Things essentially relates to Internet-connected devices, which may collect user data and conduct analytics. Recent trends and developments have however shown that the Law is still trying to evolve along the lines of the Internet of Things, and that there are diverse legal issues which have permeated and are still very much permeating the IoT space. The core issues which will be discussed in this paper are the issues of Data Privacy and Protection, Antitrust, and the determination of who bears liability when it has to do with breach of regulations in the IoT space.

This paper aims to:

1. Explain the concept of the Internet of Things
2. Identify the legal issues which are associated with the Internet of Things
3. Give a brief insight as regards how these legal issues are affecting the seamless growth of the Internet of Things across the globe
4. Shed light on how industry players are dealing with these regulatory issues 5. Giving recommendations on how to deal with these issues

BACKGROUND

According to Techtarget, the Internet of things is a system of interrelated computing devices, mechanical and digital machines, objects, animals or people that are provided with Unique identifiers (UIDs), and the ability to transfer data over a network without having to require human-to-human or human-to-computer interaction. Gillis (2022).

Techtarget further describes a 'Thing' in the Internet of Things to include a person with a heart monitor implant, a farm animal with a biochip transponder, an automobile that has built in sensors to alert the drive when tire pressure is low or any other artificial/man-made object that can be assigned an Internet Protocol (IP) address and is able to transfer data over a network. Gillis (2022).

Anirudh Sarin, in his article titled 'Legal Issues pertaining to the Internet of Things (IoT)' listed and explained the legal issues pertaining to the Internet of Things to include Data Privacy, Liability Issues, Issues as to Data Ownership, Privity of E-Contracts, Product liability and consumer protection, and issues relating to Intellectual Property rights. **Sarin (2018).**

LEGAL ISSUES AS REGARDS THE INTERNET OF THINGS

The economic impact of the Internet of Things is gigantic. It is the foundation for new innovations with a couple of benefits. IoT technologies represent $14 trillion (Bradley, Barbier, Handier, (2013) of value creation in a decade. Moreover, as a disruptive sector, the IoT opens up real opportunities for creatives who are seeking to distinguish themselves in the IoT space. In addition, IoT increases labor productivity and overall growth. Worthy of note is the fact that the full potential of the Internet of Things is yet to be discovered. In the realm of the Internet of Things, there are some legal issues with exist. It is important that these issues are dealt with in order to forestall anarchy, and to maintain law and order along the lines of the Internet of Things.

DATA PRIVACY AND SECURITY IMPLICATIONS

IoT devices deal with quantum amounts of data like Bank Account Numbers, personal preferences, Blood groups, to name a few. For example, there is no telling the extent of the risk which is associated with the hacking of a smart self-driving car. As more IoT devices emerge (and they emerge fast, based on the statistics of the growth of these devices), the potential for a data privacy breach is increased, and thus, more steps would have to be taken to stop the potential for the data breach.

Owing to the numerous benefits that the Internet of Things through IoT devices offer, IoT devices are used for more than just personal use, but also, institutional and sector-based use. For example, in hospitals, IoT devices can be used to track and monitor the vital information of patients, which in turn can be used to decipher the required medication for these given patients. While this hospital-based innovation seems to pose a number of advantages, it will be disastrous if these medical systems are hacked.

In the realms of Data Protection, Data Subjects give their consent before their data is collected. In IoT systems however, obtaining consent from data subjects and adhering to requirements of privacy proves difficult, due to the dearth of standardization processes for data security layers.

In the Global space, data privacy breach is tangibly affecting the seamless operation of the Internet of Things. With the Internet of Things, consumers let go of their privacy in bits. As new technologies are being rolled out, consumers will keep buying high-tech products that possess the innate capability to track them.

An issue had to be address where it was found that in the Samsung Smart TV privacy policy, consumers were warned not to discuss sensitive topics near the device. This met a lot of criticism and Samsung had to clarity the TV's data collection practices and edit its privacy policy.

Antitrust

Competition has been seen to be taken to the extreme when it has to do with the Internet of Things. Companies which create more IoT devices have competitive advantage against those that have fewer devices. A consumer is more likely to buy a Mac book if he owns an iPhone, because a Mac book and an iPhone would sync better together. The same goes for IoT devices.

In the area of Antitrust, voice assistant services are major exemplifiers of unhealthy competitive practices. Remote control is now a major feature which is integrated in many smart devices. Amazon uses Alexa, Google's remote control is its Google Assistant, while Apple uses Siri. Looking at the world on a global scale, there are only five voice assistant services that have gained dominance, which involve Amazon's Alexa, iPhone's Siri, Google's Google Assistant, Microsoft's Cortana, and Samsung's Bixby.

IoT providers need the services of voice assistant providers, but alas, the supply of voice assistance of limited. Lomas (2021) Some, if not all voice assistant service providers have exclusivity of their services by practices which including pre-installing or setting as default their services and voice assistants on several smart devices. Although to the disadvantage of smaller actors and players, this creates a competitive advantage for the major players, as they have access to a vast quantum of data. These voice assistants have access to, and obtain a lot of data, manage their data and user flows, and can inadvertently improve on the quality of voice assistance which they render, based on machine learning and smart algorithmic training.

Smaller players in the IoT space are invariably positioned for disadvantage because they don't have the form of access which big players have to those amounts of data. There are unhealthy competition practices which evolve based on the increasing production and interconnectedness of IoT devices.

Liability as Regards to IoT Products

As a lot of IoT devices come on to the scenes, the issue of the liability of manufacturers for certain product defects, data loss due to negligence, cannot be overemphasized. Although not seemingly obvious to users, IoT devices could cause damage in several ways, including but not limited to loss of data, loss of privacy, and identity theft. Also, like a few examples given above (Hospitals and Self-driving cars), IoT devices could cause damage to the physical well-being of the users.

An IoT device is constituted of its hardware compartment, the software compartment, as well as other service elements. This informs us of the fact that there are different levels of production involving different actors and players in the development of IoT devices. Each component and element involved in the production have their diverse warranties and disclaimers. It is usually difficult to pinpoint the particular actor on the IoT transaction chain that is responsible for a default, and once the particular liable actor cannot be triangulated, it is hard for the aggrieved person to claim any form of compensation.

For example, in a smart self-driving car, it could happen that the car registers a wrong speed limit, and starts speeding, thereby resulting in the driver being subject to some form of trouble with the road-worthy officials, or in a more serious circumstance, resulting in the death of the driver. There has been a situation where autopilot driverless technology resulted in the death of the driver. Hawkins (2021)

HOW INDUSTRY PLAYERS ARE DEALING WITH REGULATIONS

In Australia, although there are no body of rules that address the concept of the Internet of Things with specificity, the Privacy Act 1988 and the Telecommunications Act 1997 deal with privacy and cybersecurity issues. The Telecommunications Act preclude service providers from using or disclosing information about service users. The Privacy Act equally requires data collectors to disclose how personal information which is collected will be stored and used.

In an attempt to address the Data Privacy issues surrounding the Internet of Things, the Australian Government released a Code of Practice on Securing the Internet of Things for Consumers which includes, but are not limited to avoiding duplicating passwords, or the usage of weak passwords, implementing vulnerability disclosure policies, keeping software securely updated, ensuring that personal data is protected, and ensuring software integrity. Lee (2020)

The US Senate resolved recently that "the United States would have to recognize the importance of consensus-based best practices and communication among stakeholders, with the understanding that businesses play an important role in the future development of the Internet of Things. The Senate equally resolved that the Government and the IoT Industry would need to work cooperatively towards shaping regulatory standards and requirements that address a vast range of issues, including data collection rules, terms and conditions, supply chain integrity, and government policies.

There are some steps the industry is taking towards shaping regulatory standards. In August 2015, the Online Trust Alliance (OTA) issued an IoT Trust Framework, which was subsequently followed by the publication of a Revised Framework in October 2015.

In more recent times, certain states in the United States have passed legislation for IoT consumer products (California's SB – 327) and the California Consumer Privacy Act (CCPA), which came into effect in January 2020. The SB-327 mandates manufacturers and IoT creatives to give each device a preprogrammed password which users must change before the access the device for the first time. These IoT regulations instruct that the devices which manufacturers create are such devices that cannot be hacked.

As regards standardization processes for the IoT industry, more activity is being channeled towards creating a global standard for IoT products. This attempt at standardization is to achieve an optimal level of interoperability, and consistency in data outputs. Participants in the IoT industry are expected to adopt procedures and processes that are consistent with existing guidelines so as to mitigate antitrust risk in the standards process (e.g, the disclosure of standard IP rights, and up-front licensing commitments).

SOLUTIONS AND RECOMMENDATIONS

A cursory glance at the Technology sector would inform the looker that the IoT sector seen a tremendous level of advancement. Although efforts have been made to regulate the sector in some climes, a bird's eye view informs that the sector is still deregulated on the global landscape. A major recommendation which would solve some of the legal problems and complications that attend the IoT sector would be ensure the putting in place of healthy structure by the Governments of the different nations of the world.

DATA PRIVACY AND PROTECTION

The Data Privacy and Protection laws of different nations and jurisdictions have to be reviewed and be upgraded if necessary to meet up with the advancements in the development of IoT devices.

In situations where IoT devices have to connect to Internet services, or gateway platforms, these IoT devices would need to connect to Data analytic services in cloud computer settings. In this situation, the regulation of that particular jurisdiction must ensure that the cloud computing company ensures the security of the cloud platform. Diverse key point indicators, and requirement must be put in place, and cloud computing companies must be seen to have satisfied those requirements before it can be safely said that data has been truly protected.

A basic principle of Data Privacy and Protection is that the consent of Data subjects is the legal basis for the processing of those Data. There has to be regulation and a standardization process that would ensure that all IoT devices have e-user consent forms in the user interface of the IoT device to make sure that the consents of users are gotten. Also, due to the interconnectivity of different devices, different devices interact with each other at different points. Regulations must provide that at each point a new device is about to connect with a device whose consent has been granted for Data processing, a fresh consent form has to be signed by the user before the new device connects.

Regulation must also cover the issue of transfer of data across jurisdictions. The bodies responsible for the processing of personal data which is to be transferred to another jurisdiction have to satisfy already laid down legal basis for the transfer of such data, and regulations must compel them to make security measures for the protection of the data.

Antitrust

Monopolistic tendencies can only be eradicated through strong regulatory frameworks. Healthy competition poses an economic advantage to any nation or jurisdiction. We earlier looked at Voice Assistant services as the major sources of monopoly in the IoT space. Regulations could be put in place that would prevent the 'big' players in the industry from pre-installing their software on devices.

Regulatory frameworks that preclude the existence of exclusivity of services, most especially voice assistants, should be enforced. The choice of a voice assistant service that gets attached to a certain device should be left at the discretion of the user. The user could have a pool of options from which he could choose. If a user has got the prerogative of choosing whatever keypad he intends using to send messages and to operate his phone, he should also be entitled to choose from a pool of options when it has to do with voice assistance services.

Data Portability could also be encouraged, as long as data portability mechanisms are compliant with Data Protection Laws.

Although the standardization landscape of the IoT sector is advantageous, the bureaucratic nature of the landscape can prevent rapid growth, especially among budding Tech companies. The rules of membership could be relaxed and made flexible in other to encourage participation and growth of more companies which are participants in the IoT sector. In other jurisdictions, efforts could be made towards the establishment of standardization processes for the different technologies which these Tech companies intend to create.

Liability

Regulations must vividly define the roles, obligations and the rights of each of the parties (Producers and Users of the IoT devices). A clear definition of legal obligations would enable each of the parties do what is required of them. Producers of IoT devices must clearly state foreseeable occurrences, and must discuss the issues of Force Majeure. This would enable the proposed users know the pros and cons of obtaining the IoT devices. By operation of the law, parties must know 'who-bears-what-liability' when issues arise.

FURTHER RESEARCH DIRECTIONS

Worthy of note is the fact that in the future, there is the high probability for the existence of intelligent applications for smarter homes, offices, cities, transportation systems, hospitals, etc.

The application of IoT is going to be revolutionary in almost all sectors of life. It is thus important to keep up research as to how the Law converges with these future potentials.

CONCLUSION

The Internet of Things dispenses a plethora of advantages today. The interconnectivity of devices presents a seamless medium for machine-to-machine interactions, thereby decreasing the existing activities of humans. For instance, medical devices which are connected to the internet can provide information to doctors that can save the lives of their patients. At a minimum, such possibilities are very alluring.

It is however imperative to note that the framework and the ecosystem of the Internet of Things can, if not properly managed and regulated, put humans at probable risk of personal and industrial injury, loss of property and lives, and definitely, loss of privacy. This would even come to bear if the IoT devices were designed for convenience rather than security.

It would therefore be the duty of Governments of diverse nations, attorneys and citizens alike to find a balance between the advantages of the Internet of Things, and the seeming overreaching consequences of the improper creation of IoT devices.

REFERENCES

114th US Congress 1st Session. (n.d.). https://www.congress.gov/114/bills/sres110/BILLS114sres110ats.pdf

Article 2, 6, 11 and 2.12 of the Nigerian Data Protection Regulation (NPDR).

Australian Government. (2020). *Code of Practice: Securing the Internet of Things for Consumers.* https://www.homeaffairs.gov.au/reports-and-pubs/files/code-of-practice.pdf

Bradley, J., Barbier, J., & Handler, D. (2013*). Embracing the Internet of everything to capture your share of $14.4 trillion.* White Paper, Cisco.

California Senate Bill Text - SB-327 Information privacy: connected devices. (n.d.). Retrieved from https://leginfo.legislature.ca.gov/faces/billTextClient.xhtml?bill_id=201720180SB327

Gillis, A. S. (2022, March 4). *What is IOT (internet of things) and how does it work? - definition from techtarget.com.* IoT Agenda. Retrieved from https://www.techtarget.com/iotagenda/definition/Internet-of-Things-IoT

Hawkins, A. J. (2021, October 21). *'Driverless' Tesla crash in Texas wasn't actually driverless, NTSB says.* The Verge. Retrieved from https://www.theverge.com/2021/10/21/22738834/tesla-crash-texas-driver-seat-occupied-ntsb

IOT Trust Framework. Internet Society. (2021, September 28). Retrieved from https://www.internetsociety.org/iot/trust-framework/

Lee, D. (2021, March 19). *The internet of things: What it is and key legal issues.* Lawpath. Retrieved from https://lawpath.com.au/blog/the-internet-of-things-what-it-is-and-key-legal-issues

Lomas, N. (2021, June 10). *Voice AIS are raising competition concerns, EU finds*. TechCrunch. Retrieved from https://techcrunch.com/2021/06/09/voice-ais-are-raising-competition-concerns-eu-finds/

Ota releases New Internet of Things Trust Framework to address global consumer concerns. Internet Society. (2019, June 3). Retrieved from https://www.internetsociety.org/news/press-releases/2015/ota-releases-new-internet-of-things-trust-framework-to-address-global-consumer-concerns/

Privacy Act of Australia 1988. Federal Register of Legislation. https://www.legislation.gov.au/Details/C2020C00237

Sarin, A. (2018). *Legal Issues Pertaining to the Internet of Things (IoT)*. Mondaq. https://www.mondaq.com/india/privacy-protection/691560/legal-issues-pertaining-tointernet-of-things-iot

Steward, J. (2022, February 14). *The Ultimate List of Internet of Things Statistics for 2022*. Findstack. Retrieved from https://findstack.com/internet-of-things-statistics/

TechTarget. (2019, May 21). *What is California Consumer Privacy Act (CCPA)? definition from whatis.com*. SearchCompliance. Retrieved from https://searchcompliance.techtarget.com/definition/California-Consumer-Privacy-Act-CCPA#:~:text=CCPA%20Requirements&text=Buys%2C%20receives%2C%20sells%2C%20or,from%20selling%20consumers'%20personal%20information

Telecommunications Act of 1998. Federal Register of Legislation. https://www.legislation.gov.au/Details/C2020C00268

Vailshery, L. S. (2021, March 8). *Global IOT and non-IoT Connections 2010-2025*. Statista. Retrieved from https://www.statista.com/statistics/1101442/iot-number-of-connected-devicesworldwide/

What is the internet of things (IOT)? (2022). Retrieved from https://www.oracle.com/internet-of-things/what-is-iot/

ADDITIONAL READING

Oracle. (2022). *What is IoT?* https://www.oracle.com/internet-of-things/what-is-iot/

KEY TERMS AND CONDITIONS

Antitrust: A legislation against trusts or combinations. It is a body of laws which are enacted to protect trade and commerce from unlawful restraint and monopolies or unfair business practices.

Artificial Intelligence: This is the ability of a digital computer or a computer controlled robot to perform tasks which are commonly associated with humans.

Data Privacy: This is the right and ability of individuals to control how their personal information is being used.

Internet of Things: The internet of things is a system of interrelated, internet connected objects which are able to collect and transfer data over a wireless network without human intervention.

Liability: The state of being legally responsible for something.

Monopoly: This is the exclusive possession or control of the supply of, or trade in a given commodity or service.

Regulation: A rule or a directive which is made and maintained by an authority.

Conclusion

When developing IoT projects, it's critical to think about security from the beginning of the research and development process. However, due to the frequency of intrusions and the difficulty of searching for potential system vulnerabilities, guaranteeing comprehensive cybersecurity of devices, networks, and data in IoT contexts is difficult. It can be challenging to include comprehensive security measures in IoT applications. Despite running into hardware limits, incorporating security features may increase the cost and development time of a solution, which is not ideal for enterprises.

This book has provided a comprehensive, up-to-date overview of IoT techniques in IoT implementations. The literature has presented a variety of perspectives and methodologies that provide an intriguing starting point for future research in this area. While the Internet of Things (IoT) has enormous promise and, if properly implemented, may substantially benefit people, there are a number of difficulties that must be addressed in order for this technology to progress. There is yet no complete IoT-based framework that unifies all of the necessary components, technologies, and standards for IoT-building.

While the idea of combining computers, sensors, and networks to monitor and manage items has been around for decades, the "Internet of Things" is entering a new era due to the recent convergence of key technology and business trends. The Internet of Things (IoT) promises to usher in a revolutionary, completely networked "smart" society, with more intricately intertwined links between objects and their environments, as well as between objects and humans.

The idea of the Internet of Things as a pervasive array of Internet-connected gadgets could radically alter people's perceptions of what it means to be "online."

While the implications are enormous, a number of obstacles may stand in the way of this goal, specifically in the realm of security, privacy, interoperability, and standards, as well as regulatory, legal, and rights legislation, and the participation of emerging economies. The Internet of Things is a complex and growing mix of technological, social, and regulatory issues that affect a wide range of stakeholders.

The Internet of Things is already here, and it's critical to manage the concerns and maximize its benefits while minimizing its hazards.

IoT is important to the Internet Society because it represents a developing component of how individuals and organizations will engage with and incorporate the Internet and network connectivity into their personal, social, and economic life. Engaging in a comprehensive debate that pits the possibilities of IoT against its potential hazards will not lead to solutions for maximizing the advantages of IoT while limiting the risks. To plan the most effective pathways forward, informed involvement, debate, and collaboration across a range of stakeholders is required.

The Internet transformed human civilisation in the early 20th century. In fact, informational modernisation techniques have resulted in the development of computers, phones, and other devices that can independently execute a variety of programs. Then came the digital data transfer revolution. This has ushered in a new era of digital communication and networking, in which machines are joined to form enormous networks that allow programs to be accessed remotely. The virtual world was created as a result of the deployment of numerous services, such as voice communication, data transfer, and entertainment, such as television, on these computers connected to these networks.

Our community is now completely reliant on the Internet, the world's largest network and one of humanity's most amazing innovations. Designers invent and generate most of the information traffic in this network via email, the web, and other user services.

Following the digitization of information, transportation, and communication, IoT is now becoming a reality. It is based on digitized data from the real-world environment, allowing people to develop more task automation that interact with the real-world environment more effectively and efficiently.

The most difficult and ultimate aims of the digitization process have recently appeared to be ubiquitous computing, pervasive computing, and ambient intelligence. To create the so-called "smart world," where the actual and virtual worlds co-exist, automatic processes are expected to be all around humanity. Not only individuals communicate through the network, but any linked device or things involved in the processes, with or without human interaction, communicate and generate traffic in the network.

IoT has become increasingly entrenched in everyday life and is configured to respond autonomously without the need for a human interventions; it is just everywhere. The Internet of Things (IoT) is a forerunner to the smart world, as it uses ubiquitous computers and networking to simplify and supply other services, such as easy monitoring of many phenomena in our environment. Environmental and everyday goods, referred to as "things," "objects," or "machines," are improved with computing and communication technologies in the Internet of Things. They become part of the communication architecture, providing a range of services based on person-to-person, person-to-machine, machine-to-person, and machine-to-machine interactions across wired and wireless networks. These interconnected devices, objects, or things will be the next Internet or network users, generating data flow for the future Internet of Things. They will provide new services that will be delivered via the current or future Internet.

New network capabilities will be added, largely inspired by physical perception, such as detecting, discovering, sensing, choosing, actuating, and acting, as well as task automation and structuring the digital world around the real world. This will be possible due to the development of technologies such as RFID and sensors, as well as robots, nanotechnology, and other technologies. These technologies turn IoT services into an interdisciplinary field in which the majority of human senses are replicated and replaced in a digital world.

The Internet of Things is linked to a slew of technical, research, economic, and societal challenges. We have taken a more "network related view" in this book The Internet of Things, to bring together current knowledge associated with what a connected object means; what the Internet means in the IoT; the issue of standardization and governance of the IoT; and what the enabling technologies of the IoT are (the closest to the market are described in detail, primarily RFID for identifying and tracking objects, sensors for sensing the environment and actuating). Wireless communication is used in both RFID and sensor technologies.

IoT services include services designed for home networking, but they do not have the same connectivity challenges as RFID or sensors, which are small devices with limited resources (memory, CPU, and, most crucially, battery). We are not neglecting other IoT-related challenges, such as the requirement for high-performance computing, the need for even faster processing, and the limits of component physics in increasing processor speeds, etc., to deal with the billions of linked things predicted to generate network traffic. To construct ubiquitous computing and design IoT services and networking, other academic disciplines will need to collaborate and interact with the networking community.

After identifying the primary IoT-enabling technologies, concerns, and challenges, the next step is to develop the network architecture and environment to efficiently support future IoT applications as identified in this book. The future Internet's networking principles and functionality will be shaped by this. Only time will tell whether IoT services are a success! Meanwhile, society is wary of some IoT services, particularly those that propose to employ RFID technology for automated chores without a clear understanding of how to preserve a person's privacy, prevent them from being monitored, and manage any other private information data. Before such services are employed in everyday contexts, these concerns must be addressed. Other

IoT services, such as touch-a-tag apps, sensor-based monitoring services, and home networking, are also on the horizon.

Compilation of References

114th US Congress 1st Session. (n.d.). https://www.congress.gov/114/bills/sres110/BILLS114sres110ats.pdf

3rd Generation Partnership Project Organisational Partners. (2017). *Proximity-based services (ProSe), 3GPP TS 23.303 V15.0.0*. Valbonne: 3GPP Organisational Partners' Publications Offices.

3rd Generation Partnership Project Organisational Partners. (2018). *Physical layer procedures for control, (3GPP TS 38.213 version 15.2.0 Release 15)*, Valbonne: 3GPP Organisational Partners' Publications Offices, 1-101.

3rd Generation Partnership Project Organizational Partners. (2014). *Technical Specification Group Services and System Aspects; Study on architecture enhancements to support Proximity-based Services (ProSe) (Release 12)," 3GPP TR 23.703 V12.0.0 (2014-02)*. Valbonne: 3GPP Organizational Partners' Publications Offices.

Aazam, M., Zeadally, S., & Harras, K. A. (2018). Deploying fog computing in industrial internet of things and industry 4.0. *IEEE Transactions on Industrial Informatics*, *14*(99), 1. doi:10.1109/TII.2018.2855198

Abdallah, A., Mansour, M. M., & Chehab, A. (2017). A Distance-Based Power Control Scheme for D2D Communications Using Stochastic Geometry. In *Proceedings of the 2017 IEEE 86th Vehicular Technology Conference (VTC-Fall)*. Toronto: IEEE.

Abi Sen, A. A., Alawfi, I. M. M., Aloufi, H. F., Bahbouh, N. M., & Alsaawy, Y. (2021). Comparison among Cooperation, Anonymity and Cloak Area approaches for Preserving Privacy of IoT. *2021 8th International Conference on Computing for Sustainable Global Development (INDIACom)*, 413-416. doi: 10.1109/INDIACom51348.2021.00073

Abi Sen, A. A., Eassa, F. A., & Jambi, K. (2018). Preserving privacy in internet of things: A survey. *Int. J. Inf. Tecnol.*, *10*, 189–200. doi:10.1007/s41870-018-0113-4

Abomhara, M., & Køien, G. M. (2015). Cyber security and the internet of things: Vulnerabilities, threats, intruders and attacks. *Journal of Cyber Security and Mobility*, 65-88.

Abu-Elkheir, M., Hayajneh, M., & Ali, N. (2013). Data Management for the Internet of Things: Design Primitives and Solution. *Sensors (Basel)*, *13*(11), 15582–15612. doi:10.3390131115582 PMID:24240599

Aftab, Gilani, Lee, Nkenyereye, Jeong, & Song. (2019). Analysis Of Identifiers On IoT Platforms. *Digital Communications and Networks*. doi:10.1016/j.dcan.2019.05.003

Ahram, T., Sargolzaei, A., Sargolzaei, S., Daniels, J., & Amaba, B. (2017). Blockchain technology innovations. *IEEE Technology and Engineering Management Society Conference, TEMSCON 2017*, 137-141. 10.1109/TEMSCON.2017.7998367

Aich, S., Chakraborty, S., Sain, M., Lee, H. I., & Kim, H. C. (2019, February). A review on benefits of IoT integrated blockchain based supply chain management implementations across different sectors with case study. In *2019 21st international conference on advanced communication technology (ICACT)* (pp. 138-141). IEEE. 10.23919/ICACT.2019.8701910

Alagic, G., Brakerski, Z., Dulek, Y., & Schaffner, C. (2021). Impossibility of Quantum Virtual Black-Box Obfuscation of Classical Circuits. In T. Malkin & C. Peikert (Eds.), Lecture Notes in Computer Science: Vol. 12825. Advances in Cryptology – CRYPTO 2021. CRYPTO 2021. Springer. https://doi.org/10.1007/978-3-030-84242-0_18.

Alblooshi, M., Salah, K., & Alhammadi, Y. (2018). Blockchain-based Ownership Management for Medical IoT (MIoT) Devices. *2018 International Conference on Innovations in Information Technology (IIT)*, 151-156. 10.1109/INNOVATIONS.2018.8606032

Al-Dhubhani, R., & Cazalas, J. M. (2018). An adaptive geo-indistinguishability mechanism for continuous LBS queries. *Wireless Networks*, *24*(8), 3221–3239. doi:10.100711276-017-1534-x

Aljosha, J., Johanna, U., Georg, M., Voyiatzis, A. G., & Edgar, W. (2017). Lightweight Address Hopping for Defending the IPv6 IoT. *Proceedings of ARES '17*.

Alliance For Internet of Things Innovation. (2018). *Identifiers in the Internet of Things (IoT)*. https://euagenda.eu/upload/publications/identifiers-in-internet-of-things-iot.pdf

Almeida, F., Duarte Santos, J., & Augusto Monteiro, J. (2020). The challenges and opportunities in the digitalization of companies in a post-COVID-19 world. *IEEE Engineering Management Review*, *48*(3), 97–103. doi:10.1109/EMR.2020.3013206

Al-Omari, M., Rawashdeh, M., Qutaishat, F., Alshira'H, M., & Ababneh, N. (2021). An intelligent tree-based intrusion detection model for cyber security. *Journal of Network and Systems Management*, *29*(2), 20. Advance online publication. doi:10.100710922-021-09591-y

Alpár, G. B., Batten, L., Moonsamy, V., Krasnova, A., Guellier, A., & Natgunanathan, I. (2016). New directions in IoT privacy using attribute-based authentication. In *ACM International Conference on Computing Frontiers* (pp. 461-466). Como Italy: ACM. 10.1145/2903150.2911710

Alrahhal, H., Alrahhal, M. S., Jamous, R., & Jambi, K. (2020). A Symbiotic Relationship Based Leader Approach for Privacy Protection in Location Based Services. *ISPRS International Journal of Geo-Information*, *9*(6), 1–22. doi:10.3390/ijgi9060408

Alshboul, Y., Bsoul, A. A. R., & Zamil, A. L. (2021). Cybersecurity of smart home systems: Sensor identity protection. *Journal of Network and Systems Management*, 29(3), 22. Advance online publication. doi:10.100710922-021-09586-9

Ambika, N. (2021). TDSJ-IoT: Trivial Data Transmission to Sustain Energy From Reactive Jamming Attack in IoT. In Encyclopedia of Information Science and Technology, Fifth Edition (pp. 528-540). IGI Global.

Ambika, N. (2020). Tackling Jamming Attacks in IoT. In *Internet of Things (IoT)* (pp. 153–165). Springer. doi:10.1007/978-3-030-37468-6_8

An, Minh, & Kim. (2018). A Study of the Z-Wave Protocol: Implementing Your Own Smart Home Gateway. *2018 3rd International Conference on Computer and Communication Systems (ICCCS)*. 10.1109/CCOMS.2018.8463281

Ande, R., Adebisi, B., Hammoudeh, M., & Saleem, J. (2020). Internet of Things: Evolution and technologies from a security perspective. *Sustainable Cities and Society*, 54, 101728. doi:10.1016/j.scs.2019.101728

Angayarkanni, S. A., Sivakumar, R., & Ramana Rao, Y. V. (2021). Hybrid Grey Wolf: Bald Eagle search optimized support vector regression for traffic flow forecasting. *Journal of Ambient Intelligence and Humanized Computing*, 12(1), 1293–1304. doi:10.100712652-020-02182-w

Ardito, L., Petruzzelli, A. M., Panniello, U., & Garavelli, A. C. (2019). Towards industry 4.0: Mapping digital technologies for supply chain management-marketing integration. *Business Process Management Journal*, 25(2), 323–346. doi:10.1108/BPMJ-04-2017-0088

Arpan, P., & Balamuralidhar, P. (2017). *Key Factors for a Realistic Internet of Things (IoT)*. https://www.methodsandtools.com/archive/realiot.php

Arpan, P., Hemant, K. R., Samar, S., & Abhijan, B. (2018). *IoT Standardization: The Road Ahead*. https://www.intechopen.com/books/internet-of-things-technology-applications-and-standardization/iot-standardization-the-road-ahead

Arseven, M. (2021). *Standard Essential Patents and Their Role in Enabling the Internet of Things*. https://www.lexology.com/library/detail.aspx?g=b614f8c7-0d02-4dd3-869e-9c1a83e30d7e

Article 2, 6, 11 and 2.12 of the Nigerian Data Protection Regulation (NPDR).

Asghar, M. H. (2015). RFID and EPC as key technology on Internet of Things (IoT). *International Journal of Clothing Science and Technology*, 6(1).

Asiri, S., & Miri, A. (2018). A sybil resistant IoT trust model using blockchains. In *IEEE Conferences on Internet of Things, Green Computing and Communications, Cyber, Physical and Social Computing, Smart Data, Blockchain, Computer and Information Technology, iThings/GreenCom/ CPSCom/SmartData/Blockchain/CIT 2018*, 1017-1026. 10.1109/Cybermatics_2018.2018.00190

Asthana, R. (2019). *Solving Common IoT Challenges*. https://dzone.com/articles/the-most-effective-method-to-overcome-4-common-iot

Aurelio, L. C., Antonio, P., & Orazio, T. (2001). QoS management in programmable networks through mobile agents. *Microprocessors and Microsystems, 25*(2), 111-120. https://www. sciencedirect.com/science/article/pii/S0141933101001041 doi:10.1016/S0141-9331(01)00104-1

Australian Government. (2020). *Code of Practice: Securing the Internet of Things for Consumers.* https://www.homeaffairs.gov.au/reports-and-pubs/files/code-of-practice.pdf

Autenrieth, P., Lorcher, C., Pfeiffer, C., Winkens, T., & Martin, L. (2018). Current significance of IT-infrastructure enabling industry 4.0 in large companies. *2018 IEEE International Conference on Engineering, Technology and Innovation, ICE/ITMC 2018 – Proceedings.* 10.1109/ICE.2018.8436244

Bagheri, M., & Movahed, S. H. (2016, November). The effect of the Internet of Things (IoT) on education business model. In *2016 12th International Conference on Signal-Image Technology & Internet-Based Systems (SITIS)* (pp. 435-441). IEEE.

Bahashwan, A. A., Anbar, M., Abdullah, N., Al-Hadhrami, T., & Hanshi, S. M. (2021). Review on Common IoT Communication Technologies for Both Long-Range Network (LPWAN) and Short-Range Network. In F. Saeed, T. Al-Hadhrami, F. Mohammed, & E. Mohammed (Eds.), *Advances on Smart and Soft Computing. Advances in Intelligent Systems and Computing* (Vol. 1188). Springer. doi:10.1007/978-981-15-6048-4_30

Bahirat, P., He, Y., Menon, A., & Knijnenburg, B. (2018). A data-driven approach to developing IoT privacy-setting interfaces. In *23rd International Conference on Intelligent User Interfaces* (pp. 165-176). Tokyo Japan: ACM. 10.1145/3172944.3172982

Bajrami, V. (2019). *What You Need to Know About IPv6.* https://www.redhat.com/sysadmin/what-you-need-know-about-ipv6

Bandara, E., Shetty, S., & Mukkamala, R. (2021). *Casper: a blockchain-based system for efficient and secure customer credential verification. J Bank Financ Technol.* doi:10.100742786-021-00036-3

Bandyopadhyay, S., Goyal, V., Dutta, S., Pramanik, S., & Sherazi, H. H. R. (2021). Unseen to Seen by Digital Steganography. In S. Pramanik, M. M. Ghonge, R. Ravi, & K. Cengiz (Eds.), *Multidisciplinary Approach to Modern Digital Steganography.* IGI Global.

Baouya, A., Chehida, S., Cantero, M., Millet, M., Bensalem, S., & Bozga, M. (2021). Formal Modeling and Simulation of Collaborative Intelligent Robots. In *Advances in Service-Oriented and Cloud Computing. ESOCC 2020. Communications in Computer and Information Science* (Vol. 1360). Springer. doi:10.1007/978-3-030-71906-7_4

Beer, M. I., & Hassan, M. F. (2018). Adaptive security architecture for protecting RESTful web services in enterprise computing environment. *Service Oriented Computing and Applications, 12*(2), 111–121. doi:10.100711761-017-0221-1

Bello, O., Zeadally, S., & Badra, M. (2016). Network layer inter-operation of Device-to-Device communication technologies in Internet of Things (IoT). *Ad Hoc Networks, 0,* 1–11.

Belsis, P., & Pantziou, G. (2014). A *k*-anonymity privacy-preserving approach in wireless medical monitoring environments. *Personal and Ubiquitous Computing, 18*(1), 61–74. doi:10.100700779-012-0618-y

Benarous, L., & Kadri, B. (2022). Obfuscation-based location privacy-preserving scheme in cloud-enabled internet of vehicles. *Peer-to-Peer Networking and Applications, 15*(1), 461–472. doi:10.100712083-021-01233-z

Bhattacharya, A., Ghosal, A., Obaid, A. J., Krit, S., Shukla, V. K., Mandal, K., & Pramanik, S. (2021). Unsupervised Summarization Approach with Computational Statistics of Microblog Data. Methodologies and Applications of Computational Statistics for Machine Intelligence, 23-37.

Bhattacharya, S., Senapati, S., Soy, S. K., Misra, C., & Barik, R. K. (2020). Performance analysis of enhanced mist-assisted cloud computing model for healthcare system. *2020 International Conference on Computer Science, Engineering and Applications, ICCSEA 2020*. 10.1109/ICCSEA49143.2020.9132914

Bhushan, B., & Sahoo, G. (2018). Recent Advances in Attacks, Technical Challenges, Vulnerabilities and their Countermeasures in Wireless Sensor Networks. *Wireless Personal Communications, 98*, 2037–2077.

Bonomi, F., Milito, R., Zhu, J., & Addepalli, S. (2012). Fog computing and its role in the internet of things. *Proceedings of the First Edition of the MCC Workshop on Mobile Cloud Computing - MCC '12*. 10.1145/2342509.2342513

Bradley, J., Barbier, J., & Handler, D. (2013*). Embracing the Internet of everything to capture your share of $14.4 trillion*. White Paper, Cisco.

Bułat, R., & Ogiela, M. R. (2022). Personalized Cryptographic Protocols - Obfuscation Technique Based on the Qualities of the Individual. In L. Barolli, H. C. Chen, & T. Enokido (Eds.), Advances in Networked-Based Information Systems. NBiS 2021. Lecture Notes in Networks and Systems (Vol. 313). Springer. https://doi.org/10.1007/978-3-030-84913-9_19.

California Senate Bill Text - SB-327 Information privacy: connected devices. (n.d.). Retrieved from https://leginfo.legislature.ca.gov/faces/billTextClient.xhtml?bill_id=201720180SB327

Čapek, J. (2018). Cybersecurity and internet of things. *IDIMT 2018: Strategic Modeling in Management, Economy and Society - 26th Interdisciplinary Information Management Talks*, 343-349.

Caron, X., Bosua, R., Maynard, S. B., & Ahmad, A. (2016). The internet of things (IoT) and its impact on individual privacy: An australian perspective. *Computer Law & Security Review, 32*(1), 4–15. doi:10.1016/j.clsr.2015.12.001

Cetinkaya, O., & Akan, O. B. (2015). A DASH7-based power metering system. *12th Annual IEEE Consumer Communications and Networking Conference*. 10.1109/CCNC.2015.7158010

Cha, H.-J., Yang, H.-K., & Song, Y.-J. (2018). A Study on the Design of Fog Computing Architecture Using Sensor Networks. *Sensors (Basel), 18*(11), 3633. doi:10.339018113633 PMID:30373132

Chanson, M., Bogner, A., Bilgeri, D., Fleisch, E., & Wortmann, F. (2019). Blockchain for the IoT: Privacy-preserving protection of sensor data. *Journal of the Association for Information Systems*, *20*(9), 1274–1309. doi:10.17705/1jais.00567

Chen, J., Liu, C., Li, H., Li, X., & Li, S. (2016). A categorized resource sharing mechanism for device-to-device communications in cellular networks. *Mobile Information Systems*, 1-10.

Chen, S., Xu, H., Liu, D., Hu, B., & Wang, H. (2014). A vision of IoT: Applications, challenges, and opportunities with china perspective. *IEEE Internet of Things Journal*, *1*(4), 349–359. doi:10.1109/JIOT.2014.2337336

Chen, X.-Y., & Jin, Z.-G. (2012). Research on Key Technology and Applications for Internet of Things. *Physics Procedia*, *33*, 561–566. doi:10.1016/j.phpro.2012.05.104

Chen, Y.-Q., Zhou, B., Zhang, M., & Chen, C.-M. (2020). Using IoT technology for computer-integrated manufacturing systems in the semiconductor industry. *Applied Soft Computing*, *89*, 89. doi:10.1016/j.asoc.2020.106065

Cho, S., Choi, J. W., & You, C. (2013). Adaptive multi-node multiple input and multiple output (MIMO) transmission for mobile wireless multimedia sensor networks. *Sensors (Basel)*, 3382–3401.

Choudhary, S., Narayan, V., Faiz, M., & Pramanik, S. (2022). Fuzzy Approach-Based Stable Energy-Efficient AODV Routing Protocol in Mobile Ad hoc Networks. In *Software Defined Networking for Ad Hoc Networks* (pp. 125–139). EAI/Springer Innovations in Communication and Computing. doi:10.1007/978-3-030-91149-2_6

Competition Bureau Canada. (2019). *Intellectual Property Enforcement Guidelines*. Author.

Competition Commission of India. (2013). *Micromax Informatics Limited And Telefonaktiebolaget LM Ericsson*. Author.

Competition Commission of India. (2014). *In Re: Intex Technologies (India) Limited And Telefonaktiebolaget LM Ericsson (Publ)*. Author.

Competition Commission of India. (2015). *In Re: M/s Best IT World (India) Private Limited (iBall) And M/s Telefonaktiebolaget L M Ericsson (Publ)*. Author.

Croft, W. L., Sack, J. R., & Shi, W. (2021). Obfuscation of images via differential privacy: From facial images to general images. *Peer-to-Peer Networking and Applications*, *14*(3), 1705–1733. doi:10.100712083-021-01091-9

Culot, G., Fattori, F., Podrecca, M., & Sartor, M. (2019). Addressing industry 4.0 cybersecurity challenges. *IEEE Engineering Management Review*, *47*(3), 79–86. doi:10.1109/EMR.2019.2927559

Daim, T., Lai, K. K., Yalcin, H., Alsoubie, F., & Kumar, V. (2020). Forecasting technological positioning through technology knowledge redundancy: Patent citation analysis of IoT, cybersecurity, and blockchain. *Technological Forecasting and Social Change*, *161*, 120329. Advance online publication. doi:10.1016/j.techfore.2020.120329

Daniel, E., & Vasanthi, N. A. (2019). LDAP: A lightweight de-duplication and auditing protocol for secure data storage in cloud environment. *Cluster Computing, 22*(S1), 1247–1258. doi:10.100710586-017-1382-6

Datta, T., Apthorpe, N., & Feamster, N. (2018). A developer-friendly library for smart home IoT privacy-preserving traffic obfuscation. In *Workshop on IoT Security and Privacy* (pp. 43-48). Budapest Hungary: ACM. 10.1145/3229565.3229567

Dayarathna, M., Fremantle, P., Perera, S., & Suhothayan, S. (2017). Role of real-time big data processing in the internet of things. In Big data management and processing (pp. 239-262) doi:10.1201/9781315154008

De Carvalho, L. G., & Eler, M. M. (2018). *Security requirements and tests for smart toys.* doi:10.1007/978-3-319-93375-7_14

Deloitte. (2018). *The Future of Connectivity in IoT Deployments.* Author.

Dhieb, N., Ghazzai, H., Besbes, H., & Massoud, Y. (2020). Scalable and secure architecture for distributed IoT systems. *2020 IEEE Technology and Engineering Management Conference, TEMSCON 2020.* 10.1109/TEMSCON47658.2020.9140108

Dika, A., & Nowostawski, M. (2018). Security vulnerabilities in ethereum smart contracts. *Proceedings - IEEE 2018 International Congress on Cybermatics: 2018 IEEE Conferences on Internet of Things, Green Computing and Communications, Cyber, Physical and Social Computing, Smart Data, Blockchain, Computer and Information Technology, iThings/GreenCom/CPSCom/ SmartData/Blockchain/CIT 2018,* 955-962. 10.1109/Cybermatics_2018.2018.00182

Domingo-Ferrer, J., & Torra, V. (2005). Ordinal, Continuous and Heterogeneous *k*-Anonymity Through Microaggregation. *Data Mining and Knowledge Discovery, 11*(2), 195–212. doi:10.100710618-005-0007-5

Dorri, A., Kanhere, S. S., Jurdak, R., & Gauravaram, P. (2017). *Blockchain for IoT security and privacy: The case study of a smart home. In IEEE international conference on pervasive computing and communications workshops (PerCom workshops).* IEEE.

Dou, H., Chen, Y., Yang, Y., & Long, Y. (2022). A secure and efficient privacy-preserving data aggregation algorithm. *Journal of Ambient Intelligence and Humanized Computing, 13*(3), 1495–1503. doi:10.100712652-020-02801-6

Dube, D. P., & Mohanty, R. P. (2020). Towards development of a cyber-security capability maturity model. *International Journal of Business Information Systems, 34*(1), 104–127. doi:10.1504/ IJBIS.2020.106800

Dushyant, K., Muskan, G., Gupta, A., & Pramanik, S. (2022). Utilizing Machine Learning Deep Learning in Cyber security: An Innovative Approach. In M. M. Ghonge, S. Pramanik, R. Mangrulkar, & D. N. Le (Eds.), *Cyber security and Digital Forensics.* Wiley.

Dutta, S., Pramanik, S., & Bandyopadhyay, S. K. (2021). Prediction of Weight Gain during COVID-19 for Avoiding Complication in Health. *International Journal of Medical Science and Current Research*, *4*(3), 1042–1052.

Elahi, H., Munir, K., Eugeni, M., Atek, S., & Gaudenzi, P. (2020). Energy Harvesting towards Self-Powered IoT Devices. *Energies*, *13*(21), 5528. doi:10.3390/en13215528

Elbasani, E., Siriporn, P., & Choi, J. S. (2020). A Survey on RFID in Industry 4.0. In G. Kanagachidambaresan, R. Anand, E. Balasubramanian, & V. Mahima (Eds.), *Internet of Things for Industry 4.0. EAI/Springer Innovations in Communication and Computing*. Springer. doi:10.1007/978-3-030-32530-5_1

Ericsson. (2017). *5G Readiness Survey 2017: An assessment of operators' progress on the road to 5G*. Stockholm: Ericsson.

European Commission, (2017). *Communication from the Commission to the European Parliament, the Council and the European Economic and Social Committee - Setting out the E.U. approach to Standard Essential Patents*. Author.

Fafoutis, X., Tsimbalo, E., Zhao, W., Chen, H., Mellios, E., Harwin, W., Piechocki, R., & Craddock, I. (2016). BLE or IEEE 802.15.4: Which Home IoT Communication Solution is more Energy-Efficient? *EAI Endorsed Transactions on Internet of Things*, *2*(5), 1–8. doi:10.4108/eai.1-12-2016.151713

Farhan, L., Kharel, R., Kaiwartya, O., Quiroz-Castellanos, M., & Alissa, A. (2018). A concise Review on Internet of Things (IoT) – Problems, Challenges and Opportunities. *11th International Symposium on Communication Systems, Networks, and Digital Signal Processing (CSNDSP)*. 10.1109/CSNDSP.2018.8471762

Feng, D., Lu, L., Yuan-Wu, Y., Li, G. Y., Feng, Y., & Li, S. (2013). Device-to-Device Communications Underlaying Cellular Networks. *IEEE Transactions on Communications*, *61*(8), 3541–3551.

Firoiu, V., Le Boudec, J., Towsley, D., & Zhi-Li, Z. (2002). Theories and models for Internet quality of service. *Proceedings of the IEEE*, 90(9), 1565-1591. 10.1109/JPROC.2002.802002

Fodor, G., Penda, D. D., Belleschi, M., Johansson, M., & Abrardo, A. (2013). A comparative study of power control approaches for device-to-device communications. In *Proceedings of 2013 IEEE International Conference on Communications (ICC)*. Budapest: IEEE.

Foltz, C. B., & Foltz, L. (2020). Mobile users' information privacy concerns instrument and IoT. *Information and Computer Security*, *28*(3), 359–371. doi:10.1108/ICS-07-2019-0090

Foster, D., White, L., Erdil, D. C., Adams, J., Argüelles, A., Hainey, B., ... Stott, L. (2019). Toward a cloud computing learning community. *Annual Conference on Innovation and Technology in Computer Science Education, ITiCSE*, 143-155. 10.1145/3344429.3372506

Furstenau, L. B., Sott, M. K., Homrich, A. J. O., Kipper, L. M., Al Abri, A. A., Cardoso, T. F., ... Cobo, M. J. (2020). 20 years of scientific evolution of cyber security: A science mapping. *Proceedings of the International Conference on Industrial Engineering and Operations Management,* 314-325.

Gallego, B. C., & Drexl, J. (2019). *IoT Connectivity Standards: How Adaptive is the Current SEP Regulatory Framework?* (Vol. 50). IIC - International Review of Intellectual Property and Competition Law.

Gao, Y. L., Chen, X. B., Xu, G., Liu, W., Dong, M.-X., & Liu, X. (2021). A new blockchain-based personal privacy protection scheme. *Multimedia Tools and Applications, 80*(20), 30677–30690. doi:10.100711042-020-09867-6

Garg, N., & Garg, R. (2017). Energy harvesting in IoT devices: A survey. *2017 International Conference on Intelligent Sustainable Systems (ICISS)*, 10.1109/ISS1.2017.8389371

Gartenberg, C. (2019). *Huawei can't officially use microSD cards in its phones going forward.* Academic Press.

Gartenberg, C. (2019). *Intel says Apple and Qualcomm's surprise settlement pushed it to exit mobile 5G.* Academic Press.

Gay, R., Jain, A., Lin, H., & Sahai, A. (2021). Indistinguishability Obfuscation from Simple-to-State Hard Problems: New Assumptions, New Techniques, and Simplification. In A. Canteaut & F. X. Standaert (Eds.), Lecture Notes in Computer Science: Vol. 12698. Advances in Cryptology – EUROCRYPT 2021. EUROCRYPT 2021. Springer. https://doi.org/10.1007/978-3-030-77883-5_4.

Geradin, D. & Katsifis, D. (2021). *End-product- vs Component-level Licensing of Standard Essential Patents in the Internet of Things Context.* Academic Press.

Giannoutakis, K. M., Spathoulas, G., Filelis-Papadopoulos, C. K., Collen, A., Anagnostopoulos, M., Votis, K., & Nijdam, N. A. (2020). A blockchain solution for enhancing cybersecurity defence of IoT. *IEEE International Conference on Blockchain, Blockchain 2020*, 490-495. 10.1109/Blockchain50366.2020.00071

Gillis, A. S. (2022, March 4). *What is IOT (internet of things) and how does it work? - definition from techtarget.com.* IoT Agenda. Retrieved from https://www.techtarget.com/iotagenda/definition/Internet-of-Things-IoT

Gonzalez, O. (2019). *Huawei gets double bad news from S.D. Association and WiFi Alliance.* Academic Press.

Gopika, D., & Panjanathan, R. (2020). Energy-efficient routing protocols for WSN based IoT applications: A review. *Materials Today: Proceedings.* Advance online publication. doi:10.1016/j.matpr.2020.10.137

Gorog, C., & Boult, T. E. (2018). Solving global cybersecurity problems by connecting trust using blockchain. *Proceedings - IEEE 2018 International Congress on Cybermatics: 2018 IEEE Conferences on Internet of Things, Green Computing and Communications, Cyber, Physical and Social Computing, Smart Data, Blockchain, Computer and Information Technology, iThings/GreenCom/CPSCom/SmartData/Blockchain/CIT 2018*, 1425-1432. 10.1109/Cybermatics_2018.2018.00243

Graham, S. (2019). *Nokia, Daimler, Continental Ramp Up Global Patent Chess Match*. Academic Press.

Grandhi, L. S., Grandhi, S., & Wibowo, S. (2021). A security-UTAUT framework for evaluating key security determinants in smart city adoption by the Australian city councils. *Proceedings - 2021 21st ACIS International Semi-Virtual Winter Conference on Software Engineering, Artificial Intelligence, Networking and Parallel/Distributed Computing*, 17-22. 10.1109/SNPDWinter52325.2021.00013

Griffy-Brown, C., Lazarikos, D., & Chun, M. (2018). Agile business growth and cyber risk. *2018 IEEE Technology and Engineering Management Conference, TEMSCON 2018*. 10.1109/TEMSCON.2018.8488397

Griffy-Brown, C., Miller, H., Zhao, V., Lazarikos, D., & Chun, M. (2019). Emerging technologies and risk: How do we optimize enterprise risk when deploying emerging technologies? *2019 IEEE Technology and Engineering Management Conference, TEMSCON 2019*. 10.1109/TEMSCON.2019.8813743

GS1. (2014). *The GS1 EPCglobal Architecture Framework*. https://www.gs1.org/sites/default/files/docs/epc/architecture_1_6-framework 20140414.pdf

Guan, J., & Zhandry, M. (2021). Disappearing Cryptography in the Bounded Storage Model. In K. Nissim & B. Waters (Eds.), Lecture Notes in Computer Science: Vol. 13043. Theory of Cryptography. TCC 2021. Springer. https://doi.org/10.1007/978-3-030-90453-1_13.

Gubbi, J., Buyya, R., Marusic, S., & Palaniswami, M. (2013). Internet of Things (IoT): A vision, architectural elements, and future directions. *Future Generation Computer Systems*, *29*(7), 1645–1660. doi:10.1016/j.future.2013.01.010

Gupta, A., Verma, A., & Pramanik, S. (2022). Security Aspects in Advanced Image Processing Techniques for COVID-19. In S. Pramanik, A. Sharma, S. Bhatia, & D. N. Le (Eds.), *An Interdisciplinary Approach to Modern Network Security*. CRC Press.

Gupta, S., Sabitha, A. S., & Punhani, R. (2019). Cyber security threat intelligence using data mining techniques and artificial intelligence. *International Journal of Recent Technology and Engineering*, *8*(3), 6133–6140. doi:10.35940/ijrte.C5675.098319

Haider, M. I., Ali, A., & Shah, D. (2021). Block cipher's nonlinear component design by elliptic curves: an image encryption application. *Multimed Tools Appl*, *80*, 4693–4718. doi:10.1007/s11042-020-09892-5

Hameed, Khan, & Hameed. (2019). Understanding Security Requirements and Challenges in Internet of Things (IoT): A Review. *Journal of Computer Networks and Communications.*

Hawkins, A. J. (2021, October 21). *'Driverless' Tesla crash in Texas wasn't actually driverless, NTSB says.* The Verge. Retrieved from https://www.theverge.com/2021/10/21/22738834/tesla-crash-texas-driver-seat-occupied-ntsb

High Court of Delhi at New Delhi. (2015). *Telefonaktiebolaget LM Ericsson (Publ) vs Competition Commission of India and Another.* Author.

High Court of Delhi at New Delhi. (2016). *Telefonaktiebolaget LM Ericsson (Publ) vs Competition Commission of India and Another.* Author.

Hitefield, S. D., Fowler, M., & Clancy, T. C. (2018). Exploiting buffer overflow vulnerabilities in software defined radios. *Proceedings - IEEE 2018 International Congress on Cybermatics: 2018 IEEE Conferences on Internet of Things, Green Computing and Communications, Cyber, Physical and Social Computing, Smart Data, Blockchain, Computer and Information Technology, iThings/GreenCom/CPSCom/SmartData/Blockchain/CIT 2018*, 1921-1927. 10.1109/Cybermatics_2018.2018.00318

Hossain, M. F. (2013). *Traffic-driven energy efficient operational mechanisms in cellular access networks* (Doctoral dissertation). The University of Sydney, Sydney.

Huberty, M. (2015). Awaiting the Second Big Data Revolution: From Digital Noise to Value Creation. *Journal of Industry, Competition and Trade, 15*(1), 35–47. doi:10.100710842-014-0190-4

Hu, G., & Duan, X. (2019). Research on the measures of safety supervision in road transportation of dangerous goods. *Conference Proceedings of the 7th International Symposium on Project Management, ISPM 2019*, 874-879.

Husain, A., Hamed, A. R., & Wasan, A. (2020). ARP Spoofing Detection for IoT Networks Using Neural Networks. *Proceedings of the Industrial Revolution & Business Management: 11th Annual PwR Doctoral Symposium (PWRDS)*, 1-9.

Ibrahim, S., Shukla, V. K., & Bathla, R. (2020). Security enhancement in smart home management through multimodal biometric and passcode. *Proceedings of International Conference on Intelligent Engineering and Management, ICIEM 2020*, 420-424. 10.1109/ICIEM48762.2020.9160331

IERC-European Research Cluster on the Internet of Things. (2014). *Internet of Things.* Joint White Paper on Internet-of-Things Identification.

Ifzarne, S., Hafidi, I., & Idrissi, N. (2021). Secure Data Collection for Wireless Sensor Network. In M. Ben Ahmed, S. Mellouli, L. Braganca, B. Anouar Abdelhakim, & K. A. Bernadetta (Eds.), *Emerging Trends in ICT for Sustainable Development. Advances in Science, Technology & Innovation (IEREK Interdisciplinary Series for Sustainable Development).* Springer. doi:10.1007/978-3-030-53440-0_26

Institute of Electrical and Electronics Engineers. (2006). *Multi-hop relay system evaluation methodology (channel model and performance metric), IEEE C802.16j-06/013r3.* IEEE.

Iorga, M., Feldman, L., Barton, R., Martin, M. J., Goren, N., & Mahmoudi, C. (2018). *Fog Computing Conceptual Model*. doi:10.6028/NIST.SP.500-325

IOT Trust Framework. Internet Society. (2021, September 28). Retrieved from https://www.internetsociety.org/iot/trust-framework/

Islam, M. N., & Kundu, S. (2018). Preserving IoT privacy in sharing economy via smart contract. In *IEEE/ACM Third International Conference on Internet-of-Things Design and Implementation (IoTDI)* (pp. 296-297). Orlando, FL: IEEE.

ITU-T. (2017). *ITU-T X6.660-Supplement on Guidelines for Using Object Identifiers for Internet of Things*. Author.

Japan Patent Office. (2018). *Guide to Licensing negotiations involving Standard Essential Patents*. Author.

Jayasingh, R. (2022). Speckle noise removal by SORAMA segmentation in Digital Image Processing to facilitate precise robotic surgery. *International Journal of Reliable and Quality E-Healthcare*, *11*(1), 1–19. Advance online publication. doi:10.4018/IJRQEH.295083

Junglas, I. A., & Watson, R. T. (2008). Location-based services. *Communications of the ACM*, *51*(3), 65–69. doi:10.1145/1325555.1325568

K.aushik, D., Garg, M., Gupta A., & Pramanik, S., (2021). Application of Machine Learning and Deep Learning in Cyber security: An Innovative Approach. In *Cybersecurity and Digital Forensics: Challenges and Future Trends*. Wiley.

Kaur, G., Agarwal, R., & Patidar, V. (2022). Color image encryption scheme based on fractional Hartley transform and chaotic substitution–permutation. *The Visual Computer*, *38*, 1027–1050. https://doi.org/10.1007/s00371-021-02066-w

Kaur, M. J., Mishra, V. P., & Maheshwari, P. (2020). The Convergence of Digital Twin, IoT, and Machine Learning: Transforming Data into Action. In M. Farsi, A. Daneshkhah, A. Hosseinian-Far, & H. Jahankhani (Eds.), *Digital Twin Technologies and Smart Cities. Internet of Things (Technology, Communications and Computing)*. Springer. doi:10.1007/978-3-030-18732-3_1

Keane, S. (2019). *Huawei membership restored by S.D. Association*. WiFi Alliance.

Khalid, A., Sundararajan, A., Hernandez, A., & Sarwat, A. I. (2019). FACTS approach to address cybersecurity issues in electric vehicle battery systems. *2019 IEEE Technology and Engineering Management Conference, TEMSCON 2019*, 10.1109/TEMSCON.2019.8813669

Khan, M. A., & Salah, K. (2018). IoT security: Review, blockchain solutions, and open challenges. *Future Generation Computer Systems*, *82*, 395–411. doi:10.1016/j.future.2017.11.022

Khatkar, M., Kumar, K., & Kumar, B. (2020). An overview of distributed denial of service and internet of things in healthcare devices. *Proceedings of International Conference on Research, Innovation, Knowledge Management and Technology Application for Business Sustainability, INBUSH 2020*, 44-48. 10.1109/INBUSH46973.2020.9392171

Compilation of References

Kirtania, S. G., Younes, B. A., Hossain, A. R., Karacolak, T., & Sekhar, P. K. (2021). CPW-Fed Flexible Ultra-Wideband Antenna for IoT Applications. *Micromachines*, *12*(4), 453. doi:10.3390/mi12040453 PMID:33920716

Kis, M., & Singh, B. (2018). A cybersecurity case for the adoption of blockchain in the financial industry. *Proceedings - IEEE 2018 International Congress on Cybermatics: 2018 IEEE Conferences on Internet of Things, Green Computing and Communications, Cyber, Physical and Social Computing, Smart Data, Blockchain, Computer and Information Technology, Things/GreenCom/CPSCom/SmartData/Blockchain/CIT 2018*, 1491-1498. 10.1109/Cybermatics_2018.2018.00252

Korea Fair Trade Commission. (2016). Review Guidelines on Unfair Exercise of Intellectual Property Rights. Author.

Kumar, G., & Tomar, P. (2018). A Survey of IPv6 Addressing Schemes for Internet of Things. *International Journal of Hyperconnectivity and the Internet of Things*, *2*(2), 43–57. doi:10.4018/IJHIoT.2018070104

Küpper, A. (2005). *Location-based services: fundamentals and operation.* John Wiley & Sons. doi:10.1002/0470092335

Kuzlu, M., Fair, C., & Guler, O. (2021). Role of Artificial Intelligence in the Internet of Things (IoT) cybersecurity. *Discov Internet Things*, *1*, 7. doi:10.1007/s43926-020-00001-4

Laplante, P. A., & Laplante, N. (2016). The internet of things in healthcare: Potential applications and challenges. *IT Professional*, *18*(3), 2–4. doi:10.1109/MITP.2016.42

Laszewski, T., & Nauduri, P. (2012). *Migrating to the Cloud: Client/Server Migrations to the Oracle Cloud.* Elsevier. doi:10.1016/B978-1-59749-647-6.00001-6

Latif, M. N. A., Aziz, N. A. A., Hussin, N. S. N., & Aziz, Z. A. (2021). Cyber security in supply chain management: A systematic review. *Logforum*, *17*(1), 49–57. doi:10.17270/J.LOG.2021555

Leach & Mealling. (2005). *A Universally Unique IDentifier (UUID) URN Namespace.* https://www.researchgate.net/publication/215758035_A_Universally_Unique_IDentifier_UUID_URN_Namespace

Leang, B., Kim, R., & Yoo, K. (2018). Real-time transmission of secured plcs sensing data. *Proceedings - IEEE 2018 International Congress on Cybermatics: 2018 IEEE Conferences on Internet of Things, Green Computing and Communications, Cyber, Physical and Social Computing, Smart Data, Blockchain, Computer and Information Technology, iThings/GreenCom/CPSCom/SmartData/Blockchain/CIT 2018*, 931-932. 10.1109/Cybermatics_2018.2018.00177

Lee, D. (2021, March 19). *The internet of things: What it is and key legal issues.* Lawpath. Retrieved from https://lawpath.com.au/blog/the-internet-of-things-what-it-is-and-key-legal-issues

Lee, T., Kim, S., & Kim, K. (2019). A research on the vulnerabilities of PLC using search engine. *ICTC 2019 - 10th International Conference on ICT Convergence: ICT Convergence Leading the Autonomous Future*, 184-188. 10.1109/ICTC46691.2019.8939961

Lin, X., Andrews, J., Ghosh, A., & Ratasuk, R. (2014). An overview of 3GPP device-to-device proximity services. *IEEE Communications Magazine, 52*(4), 40–48.

Lionel, S. V. (n.d.). *Number of internet of things (IoT) connected devices worldwide in 2018, 2025 and 2030.* https://www.statista.com/statistics/802690/worldwide-connected-devices-by-access-technology/

Liu, C. H., Yang, B., & Liu, T. (2014). Efficient Naming, Addressing and Profile Services In Internet-of-Things Sensory Environments. *Ad Hoc Networks, 18*, 18. doi:10.1016/j.adhoc.2013.02.008

Liu, C., Feng, Y., Lin, D., Wu, L., & Guo, M. (2020). Iot based laundry services: An application of big data analytics, intelligent logistics management, and machine learning techniques. *International Journal of Production Research, 58*(17), 5113–5131. doi:10.1080/00207543.2019.1677961

Liu, Y., & Zhao, Q. (2019). E-voting scheme using secret sharing and K-anonymity. *World Wide Web (Bussum), 22*(4), 1657–1667. doi:10.100711280-018-0575-0

Lomas, N. (2021, June 10). *Voice AIS are raising competition concerns, EU finds.* TechCrunch. Retrieved from https://techcrunch.com/2021/06/09/voice-ais-are-raising-competition-concerns-eu-finds/

Lopez, S. M. (2016). *An overview of D2D in 3GPP LTE standard.* Retrieved from http://d2d-4-5g.gforge.inria.fr/Workshop-June2016/slides/Overview_LTE_D2D.pdf

Luh, R., Marschalek, S., Kaiser, M., Janicke, H., & Schrittwieser, S. (2017). Semantics-aware detection of targeted attacks: A survey. *J Comput Virol Hack Tech, 13*(1), 47–85. doi:10.100711416-016-0273-3

Luo, J. N., Yang, M. H., & Tsai, K. Y. (2019). A geographic map-based middleware framework to obfuscate smart vehicles' locations. *Multimed Tools Appl, 78*, 28877–28902. doi:10.1007/s11042-019-7350-9

Luo, B., & Sun, Z. (2015). Research on the Model of a Lightweight Resource Addressing. *Chinese Journal of Electronics, 24*(4), 832–836. doi:10.1049/cje.2015.10.028

Ma, Y., Li, Y., Zhang, Z., Zhang, R., Liu, L., & Zhang, X. (2021). A Classic Multi-method Collaborative Obfuscation Strategy. In Y. Tan, Y. Shi, A. Zomaya, H. Yan, & J. Cai (Eds.), Data Mining and Big Data. DMBD 2021. Communications in Computer and Information Science (Vol. 1454). Springer. https://doi.org/10.1007/978-981-16-7502-7_10

Mach, P., Becvar, Z., & Vanek, T. (2015). In-band device-to-device communication in OFDMA cellular Networks: A Survey and Challenges. *IEEE Communications Surveys and Tutorials, 17*(4), 1885–1922.

Mahdi, A. H. (2016). *The integration of device-to-device communication in future cellular systems* (Doctoral dissertation). Technischen Universität, Berlin.

Ma, L., & Wang, Y. (2011). Research on formation mechanism of information risk in supply chain and its control countermeasures. *International Conference on Management and Service Science, MASS 2011*, 10.1109/ICMSS.2011.5998265

Mandal, A., Dutta, S., & Pramanik, S. (2021). Machine Intelligence of Pi from Geometrical Figures with variable Parameters usingSCILab. In *Methodologies and Applications of Computational Statistics for Machine Intelligence*. IGI Global. doi:10.4018/978-1-7998-7701-1.ch003

Mandrakov, E. S., Vasiliev, V. A., & Dudina, D. A. (2020). Non-conforming products management in a digital quality management system. *Proceedings of the 2020 IEEE International Conference "Quality Management, Transport and Information Security, Information Technologies", IT and QM and IS 2020*, 266-268. 10.1109/ITQMIS51053.2020.9322931

Mangino, A., Pour, M. S., & Bou-Harb, E. (2020). Internet-scale insecurity of consumer internet of things. *ACM Transactions on Management Information Systems, 11*(4), 1–24. Advance online publication. doi:10.1145/3394504

Manoj Kumar, N., & Dash, A. (2017, November). Internet of things: an opportunity for transportation and logistics. *Proceedings of the International Conference on Inventive Computing and Informatics (ICICI 2017)*, 194-197.

Martynov, V. V., Shavaleeva, D. N., & Zaytseva, A. A. (2019). Information technology as the basis for transformation into a digital society and industry 5.0. *Proceedings of the 2019 IEEE International Conference Quality Management, Transport and Information Security, Information Technologies IT and QM and IS 2019*, 539-543. 10.1109/ITQMIS.2019.8928305

Mavropoulos, O., Mouratidis, H., Fish, A., Panaousis, E., & Kalloniatis, C. (2016). *APPARATUS: Reasoning about security requirements in the internet of things*. doi:10.1007/978-3-319-39564-7_21

McDonagh, L., & Bonadio, E. (2019). *Standard Essential Patents and the Internet of Things: In-Depth Analysis*. Academic Press.

MediaTek. (2018). *5G NR Uplink enhancements better cell coverage & user experience*. Author.

Mell & Grance. (2009). *A NIST Notional Definition of Cloud Computing, version 15*. NIST.

Mickle, T., Kendall, B., & Fitch, A. (2019). *Qualcomm's Practices Violate Antitrust Law*. Judge Rules.

Miorandi, D., Sicari, S., De Pellegrini, F., & Chlamtac, I. (2012). Internet of Things: Vision, Applications and Research Challenges. *Ad Hoc Networks, 10*(7), 1497–1516. doi:10.1016/j.adhoc.2012.02.016

Mohamed, L., Nabil, K., Khalid, M., Abdellah, E., & Mohamed, F. (2020). IoT security: Challenges and countermeasures. *Procedia Computer Science, 177*, 503-508. https://www.sciencedirect.com/science/article/pii/S1877050920323395 doi:10.1016/j.procs.2020.10.069

Mohamed, K. S. (2019). IoT Networking and Communication Layer. In *The Era of Internet of Things* (pp. 49–70). Springer. doi:10.1007/978-3-030-18133-8_3

Mohanty, S., & Vyas, S. (2018). How to compete in the age of artificial intelligence: Implementing a collaborative human-machine strategy for your business. *How to compete in the age of artificial intelligence: Implementing a collaborative human-machine strategy for your business.* doi:10.1007/978-1-4842-3808-0

Morris, D., Madzudzo, G., & Garcia-Perez, A. (2018). Cybersecurity and the auto industry: The growing challenges presented by connected cars. *International Journal of Automotive Technology and Management, 18*(2), 105–118. doi:10.1504/IJATM.2018.092187

Mueller, F. (2021a). *Sisvel becomes third Avanci licensor to sue Ford Motor Company over cellular standard-essential patents.* Academic Press.

Mueller, F. (2021b). *L2 Mobile Technologies claims Qualcomm chips in Ford, Lincoln cars infringe 3G standard-essential patents originally obtained by ASUSTeK.* Academic Press.

Mueller, F. (2021c). *Japanese patent licensing firm I.P. Bridge is suing Ford Motor Company in Munich over former Panasonic SEP.* Academic Press.

Munhoz, A. M., Chala, L., & Melo, G. d. (2021). Clinical and MRI Evaluation of Silicone Gel Implants with RFID-M Traceability System: A Prospective Controlled Cohort Study Related to Safety and Image Quality in MRI Follow-Up. *Aesthetic Plastic Surgery, 45*, 2645–2655. https://doi.org/10.1007/s00266-021-02355-8

Mwashita, W., & Odhiambo, M. O. (2018). Interference management techniques for device-to-device communications. In P. K. Gupta, T. I. Ören, & M. Singh (Eds.), Predictive Intelligence Using Big Data and the Internet of Things (pp. 219-245). IGI Global.

Nadeem, M. (2021). Board Gender Diversity and Managerial Obfuscation: Evidence from the Readability of Narrative Disclosure in 10-K Reports. *Journal of Business Ethics.* Advance online publication. doi:10.100710551-021-04830-3

Naeini, P. E., Bhagavatula, S., Habib, H., Degeling, M., Bauer, L., Cranor, L. F., & Sadeh, N. (2017). Privacy expectations and preferences in an IoT world. In *Thirteenth Symposium on Usable Privacy and Security ({SOUPS})* (pp. 399-412). Santa Clara, CA: USENIX.

Nagaraj, A. (2021). Introduction to Sensors in IoT and Cloud Computing Applications. Bentham Science Publishers. doi:10.2174/97898114793591210101

Nakai, T., Misawa, Y., & Tokushige, Y. (2021). How to Solve Millionaires' Problem with Two Kinds of Cards. *New Generation Computing, 39*, 73–96. https://doi.org/10.1007/s00354-020-00118-8-m

Nash, I. (2021). Cybersecurity in a post-data environment: Considerations on the regulation of code and the role of producer and consumer liability in smart devices. *Computer Law & Security Review, 40*, 105529. Advance online publication. doi:10.1016/j.clsr.2021.105529

Compilation of References

Nekrasov, H. A., & Polivoda, D. E. (2019). The development of a reference architectural prototype of the internet of things network mode. *Proceedings of the 2019 IEEE International Conference Quality Management, Transport and Information Security, Information Technologies IT and QM and IS 2019*, 554-557. 10.1109/ITQMIS.2019.8928303

Nižetić, S., Šolić, P., López-de-Ipiña González-de-Artaza, D., & Patrono, L. (2020). Internet of Things (IoT): Opportunities, issues, and challenges towards a smart and sustainable future. *Journal of Cleaner Production*, *274*, 122877. doi:10.1016/j.jclepro.2020.122877 PMID:32834567

Oconnor, T. J., & Stricklan, C. (2021). Teaching a hands-on mobile and wireless cybersecurity course. *Annual Conference on Innovation and Technology in Computer Science Education, ITiCSE*, 296-302. 10.1145/3430665.3456346

Odhiambo Marcel, O., & Umenne, P. O. (2012). Net-Computer: Internet Computer Architecture and its Application in E-Commerce. *Journal: Engineering, Technology & Applied Science Research*, *2*(6), 302 – 309.

Olive, C. (2012). *White Paper: Cloud Computing Characteristics are Key*. https://www.gpstrategies.com/wpcontent/uploads/2016/04/wpCloudCharacteristics.pdf

Oracle India. (2020). *What is IoT?* Author.

Oravec, J. A. (2017). Kill switches, remote deletion, and intelligent agents: Framing everyday household cybersecurity in the internet of things. *Technology in Society*, *51*, 189–198. doi:10.1016/j.techsoc.2017.09.004

Ota releases New Internet of Things Trust Framework to address global consumer concerns. Internet Society. (2019, June 3). Retrieved from https://www.internetsociety.org/news/press-releases/2015/ota-releases-new-internet-of-things-trust-framework-to-address-global-consumer-concerns/

Pandey, B. K., Pandey, D., Wairya, S., Agarwal, G., Dadeech, P., Dogiwal, S. R., & Pramanik, S. (2022). Application of Integrated Steganography and Image Compressing Techniques for Confidential Information Transmission. In Cyber Security and Network Security. Wiley. doi:10.1002/9781119812555.ch8

Pandey, P., & Litoriya, R. (2021). Securing E-health Networks from Counterfeit Medicine Penetration Using Blockchain. *Wireless Personal Communications*, *117*, 7–25. https://doi.org/10.1007/s11277-020-07041-7

Parasol, M. (2018). The impact of china's 2016 cyber security law on foreign technology firms, and on china's big data and smart city dreams. *Computer Law & Security Review*, *34*(1), 67–98. doi:10.1016/j.clsr.2017.05.022

Paul, J., & Criado, A. R. (2020). The art of writing literature review: What do we know and what do we need to know? *International Business Review*, *29*(4), 101717. doi:10.1016/j.ibusrev.2020.101717

Petrie, H., Johnson, V., Strothotte, T., Raab, A., Fritz, S., & Michel, R. (1996). MoBIC: Designing a travel aid for blind and elderly people. *Journal of Navigation, 49*(1), 45–52. doi:10.1017/S0373463300013084

Podszun, R. (2019). Standard Essential Patents and Antitrust Law in the Age of Standardisation and the Internet of Things: Shifting Paradigms. *IIC - International Review of Intellectual Property and Competition Law, 50*, 720-745.

Pokhrel, S. R., Vu, H. L., & Cricenti, A. L. (2019). Adaptive admission control for IoT applications in home WiFi networks. *IEEE Transactions on Mobile Computing, 19*(12), 2731–2742. doi:10.1109/TMC.2019.2935719

Poluru, R. K., & Naseera, S. (2017). A Literature Review on Routing Strategy in the Internet of Things. *Journal of Engineering Science and Technology Review, 10*(5), 50–60. doi:10.25103/jestr.105.06

Porter, J. (2019). *Apple will try to tear apart Qualcomm's biggest business in court this week.* Academic Press.

Postscapes and Harbor Research. (2015). *What Exactly is the Internet of Things.* https://www.visualistan.com/2015/09/what-exactly-is-internet-of-things.html

Pradeep, Kousalya, Suresh, & Edwin. (2016). IoT And Its Connectivity Challenges In Smart Home. *International Research Journal of Engineering and Technology, 3*(12).

Pradhan, D., Sahu, P. K., Goje, N. S., Ghonge, M. M., & Pramanik, S. (2022). Security, Privacy, Risk, and Safety Toward 5G Green Network (5G-GN). In Cyber Security and Network Security. Wiley. doi:10.1002/9781119812555.ch9

Pramanik, S., & Raja, S. (2019). Analytical Study on Security Issues in Steganography. *Think-India, 22*(35), 106-114.

Pramanik, S., Joardar, S., Jena, O. P., & Obaid, A. J. (2022). An Analysis of the Operations and Confrontations of Using Green IT in Sustainable Farming. *Al-Kadhum 2nd International Conference on Modern Applications of Information and Communication Technology.*

Pramanik, S., Pandey, D., Joardar, S., Niranjanamurthy, M., Pandey, B. K., & Kaur, J. (2022). An Overview of IoT Privacy and Security in Smart Cities. *ICCAPE 2022.*

Pramanik, S., Sagayam, K. M., & Jena, O. P. (2021). Machine Learning Frameworks in Cancer Detection. ICCSRE 2021.

Pramanik, S., & Bandyopadhyay, S. K. (2013). Application of Steganography in Symmetric Key Cryptography with Genetic Algorithm. *International Journals of Engineering and Technology, 10*, 1791–1799.

Pramanik, S., & Bandyopadhyay, S. K. (2014). *An Innovative Approach in Steganography.* Scholars Journal of Engineering and Technology.

Pramanik, S., & Bandyopadhyay, S. K. (2014). Image Steganography Using Wavelet Transform and Genetic Algorithm. *International Journal of Innovative Research in Advanced Engineering*, *1*, 1–4.

Pramanik, S., & Raja, S. (2020). A Secured Image Steganography using Genetic Algorithm. *Advances in Mathematics: Scientific Journal*, *9*(7), 4533–4541.

Privacy Act of Australia 1988. Federal Register of Legislation. https://www.legislation.gov.au/Details/C2020C00237

Purkayastha, K. D., Mishra, R. K., Shil, A., & Pradhan, S. N. (2021). IoT Based Design of Air Quality Monitoring System Web Server for Android Platform. *Wireless Personal Communications*, *118*(4), 2921–2940. doi:10.100711277-021-08162-3

Purri, S., Choudhury, T., Kashyap, N., & Kumar, P. (2017). Specialization of IoT applications in health care industries. In *International Conference on Big Data Analytics and Computational Intelligence (ICBDAC)* (pp. 252-256). Chirala, India: IEEE. 10.1109/ICBDACI.2017.8070843

Puthal, Sahoo, Mishra, & Swain. (2015). Cloud Computing Features, Issues and Challenges: A Big Picture. *International Conference on Computational Intelligence & Networks*.

Raimundo, R., & Rosário, A. (2021). Blockchain system in the Higher Education. *European Journal of Investigation in Health, Psychology and Education, 11*(1), 276-293. doi:10.3390/ejihpe11101002

Rajashree, S., Gajkumar Shah, P., & Murali, S. (2018). Security model for internet of things end devices. *Proceedings - IEEE 2018 International Congress on Cybermatics: 2018 IEEE Conferences on Internet of Things, Green Computing and Communications, Cyber, Physical and Social Computing, Smart Data, Blockchain, Computer and Information Technology, iThings/GreenCom/CPSCom/SmartData/Blockchain/CIT 2018, 219-221.* 10.1109/Cybermatics_2018.2018.00066

Ranger, S. (2020). *What is the IoT? Everything you need to know about the Internet of Things right now.* https://www.zdnet.com/article/what-is-the-internet-of-things-everything-you-need-to-know-about-the-iot-right-now/

Rappaport, T. S., Xing, Y., MacCartney, G. R., Molisch, A. F., Mellios, E., & Zhang, J. (2017). Overview of Millimetre Wave Communications for Fifth Generation (5G) Wireless Networks-With a Focus on Propagation Models. *IEEE Transactions on Antennas and Propagation, 65*(12), 6213–6230.

Rathee, D., Ahuja, K.& Nayyar, A. (2019). Sustainable future IoT services with touch-enabled handheld devices. *Security and Privacy of Electronic Healthcare Records: Concepts, paradigms and solutions,* 131-152.

Rhahla, M., Abdellatif, T., Attia, R., & Berrayana, W. (2019). A GDPR controller for IoT systems: Application to e-health. *Proceedings - 2019 IEEE 28th International Conference on Enabling Technologies: Infrastructure for Collaborative Enterprises, WETICE 2019, 170-173.* 10.1109/WETICE.2019.00044

Rosário, A. (2021). Research-Based Guidelines for Marketing Information Systems. *International Journal of Business Strategy and Automation, 2*(1), 1–16. doi:10.4018/IJBSA.20210101.oa1

Rosário, A., & Cruz, R. (2019). Determinants of Innovation in Digital Marketing, Innovation Policy and Trends in the Digital Age. *Journal of Reviews on Global Economics, 8*, 1722–1731. doi:10.6000/1929-7092.2019.08.154

Rosário, A., Fernandes, F., Raimundo, R., & Cruz, R. (2021). Determinants of Nascent Entrepreneurship Development. In A. Carrizo Moreira & J. G. Dantas (Eds.), *Handbook of Research on Nascent Entrepreneurship and Creating New Ventures* (pp. 172–193). IGI Global. doi:10.4018/978-1-7998-4826-4.ch008

Rountree, D., & Castrillo, I. (2014). *Cloud Deployment Models, The Basics of Cloud Computing.* Elsevier.

Safdar, G. A., Ur-Rehman, M., Muhammad, M., Imran, M. A., & Tafazolli, R. (2016). Interference Mitigation in D2D Communication Underlaying LTE-A Network. *IEEE Access: Practical Innovations, Open Solutions, 4*, 7967–7987.

Sahoo, K. C., & Pati, U. C. (2020). IoT based intrusion detection system using PIR sensor. *2017 2nd IEEE International Conference on Recent Trends in Electronics, Information & Communication Technology (RTEICT)*, 1641-1645. 10.1109/RTEICT.2017.8256877

Saksonov, E. A., Leokhin, Y. L., & Azarov, V. N. (2019). Organization of information security in industrial internet of things systems. *2019 IEEE International Conference Quality Management, Transport and Information Security, Information Technologies IT and QM and IS 2019*, 3-7. 10.1109/ITQMIS.2019.8928442

Samaila, M. G., Neto, M., Fernandes, D. A. B., Freire, M. M., & Inácio, P. R. M. (2018). Challenges of securing Internet of Things devices: A survey. *Security and Privacy, 1*(2), e20. doi:10.1002py2.20

Samanta, D., Dutta, S., Galety, M. G., & Pramanik, S. (2021). A Novel Approach for Web Mining Taxonomy for High-Performance Computing. *The 4th International Conference of Computer Science and Renewable Energies (ICCSRE'2021)*. 10.1051/e3sconf/202129701073

Samuel, S. S. I. (2016). A Review Of Connectivity Challenges In Iot-Smart Home. *2016 3rd MEC International Conference on Big Data and Smart City (ICBDSC)*. 10.1109/ICBDSC.2016.7460395

Sani, A. S., Yuan, D., Jin, J., Gao, L., Yu, S., & Dong, Z. Y. (2019). Cyber security framework for Internet of Things-based Energy Internet. *Future Generation Computer Systems, 93*, 849–859. doi:10.1016/j.future.2018.01.029

Sanzgiri, A., & Dasgupta, D. Classification of insider threat detection techniques. In *Proceedings of the 11th Annual Cyber and Information Security Research Conference.* ACM.

Sarin, A. (2018). *Legal Issues Pertaining to the Internet of Things (IoT).* Mondaq. https://www.mondaq.com/india/privacy-protection/691560/legal-issues-pertaining-tointernet-of-things-iot

Sarı, T., Güleş, H. K., & Yiğitol, B. (2020). Awareness and readiness of industry 4.0: The case of Turkish manufacturing industry. *Advances in Production Engineering & Management, 15*(1), 57–68. doi:10.14743/apem2020.1.349

Sarkar, M., Pramanik, S., & Sahnwaj, S. (2020). Image Steganography Using DES-DCT Technique. *Turkish Journal of Computer and Mathematics Education, 11*(2), 906–913.

Sasikala, P. (2013). Research Challenges And Potential Green Technological Applications In Cloud Computing. *International Journal of Cloud Computing, 2*(1), 1–19. doi:10.1504/IJCC.2013.050953

Sen, A. A. A., & Yamin, M. (2021). Advantages of using fog in IoT applications. *Int. J. Inf. Tecnol., 13*, 829–837. doi:10.1007/s41870-020-00514-9

Sethi, P., & Sarangi, S. R. (2017). Internet of Things: Architectures, Protocols, and Applications. *Journal of Electrical and Computer Engineering, 2017*, 9324035. doi:10.1155/2017/9324035

Setiawan, A. B., Syamsudin, A., & Sastrosubroto, A. S. (2017). Information security governance on national cyber physical systems. *2016 International Conference on Information Technology Systems and Innovation, ICITSI 2016 - Proceedings*, 10.1109/ICITSI.2016.7858210

Shacklett, M. (2021). *4 IoT Connectivity Challenges and Strategies to Tackle Them*. https://internetofthingsagenda.techtarget.com/feature/4-IoT-connectivity-challenges-and-strategies-to-tackle-them

Shankar, K., Elhoseny, M., Kumar, R. S., Lakshmanaprabu, S. K., & Yuan, X. (2020). Secret image sharing scheme with encrypted shadow images using optimal homomorphic encryption technique. *Journal of Ambient Intelligence and Humanized Computing, 11*(5), 1821–1833. doi:10.100712652-018-1161-0

Shapiro, C. (2001). Navigating the Patent Thicket: Cross Licenses, Patent Pools, and Standard Setting. In A. B. Jaffe, J. Lerner, & S. Stern (Eds.), *Innovation Policy and the Economy* (pp. 119–150). MIT Press.

Sharma, S. K., & Ahmed, S. S. (2021). IoT-based analysis for controlling & spreading prediction of COVID-19 in Saudi Arabia. *Soft Computing, 25*(18), 12551–12563. doi:10.100700500-021-06024-5 PMID:34305445

Sharma, & Vijay, & Shukla. (2020). Ultra-Wideband Technology: Standards, Characteristics, Applications. *Helix, 10*(4), 59–65. doi:10.29042/2020-10-4-59-65

Siddiqie, S., Mondal, A., & Reddy, P. K. (2021). An Improved Dummy Generation Approach for Enhancing User Location Privacy. In Lecture Notes in Computer Science: Vol. 12683. Database Systems for Advanced Applications. DASFAA 2021. Springer. https://doi.org/10.1007/978-3-030-73200-4_33.

Siddiqui, I. F., Qureshi, N. M. F., Chowdhry, B. S., & Uqaili, M. A. (2020). Pseudo-Cache-Based IoT Small Files Management Framework in HDFS Cluster. *Wireless Personal Communications, 113*(3), 1495–1522. doi:10.100711277-020-07312-3

Silverajan, B., Ocak, M., & Nagel, B. (2018). Cybersecurity attacks and defences for unmanned smart ships. *2018 International Congress on Cybermatics: 2018 IEEE Conferences on Internet of Things, Green Computing and Communications, Cyber, Physical and Social Computing, Smart Data, Blockchain, Computer and Information Technology, iThings/GreenCom/CPSCom/ SmartData/Blockchain/CIT 2018*, 15-20. 10.1109/Cybermatics_2018.2018.00037

Sinha, M., Chacko, E., Makhija, P., & Pramanik, S. (2021). Energy Efficient Smart Cities with Green IoT. In *Green Technological Innovation for Sustainable Smart Societies: Post Pandemic Era, C. Chakrabarty*. Springer.

Sinha, M., Chacko, E., Makhija, P., & Pramanik, S. (2021). Energy-Efficient Smart Cities with Green Internet of Things. In C. Chakraborty (Ed.), *Green Technological Innovation for Sustainable Smart Societies*. Springer. doi:10.1007/978-3-030-73295-0_16

Sivakumar, S., Siddappa Naidu, K., & Karunanithi, K. (2019). Design of energy management system using autonomous hybrid micro-grid under IOT environment. *International Journal of Recent Technology and Engineering, 8*(2), 338-343. doi:10.35940/ijrte.B1058.0782S219

Smith, K. J., Dhillon, G., & Carter, L. (2021). User values and the development of a cybersecurity public policy for the IoT. *International Journal of Information Management, 56*, 102123. Advance online publication. doi:10.1016/j.ijinfomgt.2020.102123

Sohrabi Safa, N., Maple, C., & Watson, T. (2017). An information security risk management model for smart industries. *Advances in Transdisciplinary Engineering, 6*, 257-262. 10.3233/978-1-61499-792-4-257

Soldani, D. (2020). On Australia's cyber and critical technology international engagement strategy towards 6G how Australia may become a leader in cyberspace. *Journal of Telecommunications and the Digital Economy, 8*(4), 127–158. doi:10.18080/jtde.v8n4.340

Soltani, R., Nguyen, U. T., & An, A. (2018). A new approach to client onboarding using self-sovereign identity and distributed ledger. *Proceedings - IEEE 2018 International Congress on Cybermatics: 2018 IEEE Conferences on Internet of Things, Green Computing and Communications, Cyber, Physical and Social Computing, Smart Data, Blockchain, Computer and Information Technology, iThings/GreenCom/CPSCom/SmartData/Blockchain/CIT 2018*, 1129-1136. 10.1109/Cybermatics_2018.2018.00205

Song, L., Niyato, D., Han, Z., & Hossain, E. (2015). *Wireless Device-to-Device Communications and Networks*. Cambridge University Press.

Stathaki, C., Xenakis, A., Skayannis, P., & Stamoulis, G. (2020). Studying the role of proximity in advancing innovation partnerships at the dawn of industry 4.0 era. *Proceedings of the European Conference on Innovation and Entrepreneurship, ECIE*, 651-658. doi:10.34190/EIE.20.048

Stephen Dass, A., & Prabhu, J. (2020). Hybrid coherent encryption scheme for multimedia big data management using cryptographic encryption methods. *International Journal of Grid and Utility Computing, 11*(4), 496–508. doi:10.1504/IJGUC.2020.108449

Steward, J. (2022, February 14). *The Ultimate List of Internet of Things Statistics for 2022*. Findstack. Retrieved from https://findstack.com/internet-of-things-statistics/

Subrahmanyam, V., Zubair, M. A., Kumar, A., & Rajalakshmi, P. (2018). A Low Power Minimal Error IEEE 802.15.4 Transceiver for Heart Monitoring in IoT Applications. *Wireless Personal Communications*, *100*(2), 611–629. doi:10.100711277-018-5255-y

Sun, S., Theodore, S. R., Sundeep, R., Timothy, A. T., Amitava, G., Istvan, Z., ... Jan, J. (2016). Propagation Path Loss Models for 5G Urban Micro- and Macro-Cellular Scenarios. In *Proceedings of the IEEE 83rd Vehicular Technology Conference (VTC Spring)*. IEEE.

Sun, G., Chang, V., Ramachandran, M., Sun, Z., Li, G., Yu, H., & Liao, D. (2017). Efficient location privacy algorithm for Internet of Things (IoT) services and applications. *Journal of Network and Computer Applications*, *89*, 3–13. doi:10.1016/j.jnca.2016.10.011

Sun, G., Liao, D., Li, H., Yu, H., & Chang, V. (2017). L2P2: A location-label based approach for privacy preserving in LBS. *Future Generation Computer Systems*, *74*, 375–384. doi:10.1016/j.future.2016.08.023

Suzuki, T., Takeshita, J. I., Ogawa, M., Lu, X.-N., & Ojima, Y. (2020). *Analysis of measurement precision experiment with categorical variables*. Paper presented at 13th International Workshop on Intelligent Statistical Quality Control 2019, IWISQC 2019, Hong Kong, Hong Kong.

Talaat, F. M., Saraya, M. S., Saleh, A. I., Ali, H. A., & Ali, S. H. (2020). A load balancing and optimization strategy (LBOS) using reinforcement learning in fog computing environment. *Journal of Ambient Intelligence and Humanized Computing*, *11*(11), 4951–4966. doi:10.100712652-020-01768-8

Talend. (n.d.). *What is Data Integrity and Why Is It Important?* https://www.talend.com/resources/what-is-data-integrity/

TechTarget. (2019, May 21). *What is California Consumer Privacy Act (CCPA)? definition from whatis.com*. SearchCompliance. Retrieved from https://searchcompliance.techtarget.com/definition/California-Consumer-Privacy-Act-CCPA#:~:text=CCPA%20Requirements&text=Buys%2C%20receives%2C%20sells%2C%20or,from%20selling%20consumers'%20personal%20information

Telecommunications Act of 1998. Federal Register of Legislation. https://www.legislation.gov.au/Details/C2020C00268

Terruggia, R., & Garrone, F. (2020). Secure IoT and cloud based infrastructure for the monitoring of power consumption and asset control. *12th AEIT International Annual Conference, AEIT 2020*. 10.23919/AEIT50178.2020.9241195

The Japan Fair Trade Commission. (2016). *Guidelines for the Use of Intellectual Property under the Antimonopoly Act*. Author.

Tomatis, A., Cataldi, P., Pau, G., Mulassano, P., & Dovis, F. (2008). Cooperative LBS for Secure Transport System. *21st International Technical Meeting of the Satellite Division of The Institute of Navigation (ION GNSS 2008)*, 861-866.

U.S. Department of Justice, Federal Trade Commission. (2017). *Antitrust Guidelines for the Licensing of Intellectual Property.* Author.

U.S. Department of Justice, U.S. Patent & Trademark Office. (2013). *Policy Statement on Remedies for Standards-Essential Patents Subject to Voluntary F/RAND Commitments.* Author.

U.S. Patent & Trademark Office, U.S. Department of Justice, & National Institute of Standards and Technology. (2019). *Policy Statement on Remedies for Standards-Essential Patents Subject to Voluntary F/RAND Commitments.* Author.

Ukil, A., Bandyopadhyay, S., & Pal, A. (2014). IoT-privacy: To be private or not to be private. In *IEEE Conference on Computer Communications Workshops (INFOCOM WKSHPS)* (pp. 123-124). Toronto, Canada: IEEE. 10.1109/INFCOMW.2014.6849186

Ullah, F., Naeem, H., Jabbar, S., Khalid, S., Latif, M. A., Al-Turjman, F., & Mostarda, L. (2019). Cyber security threats detection in internet of things using deep learning approach. *IEEE Access: Practical Innovations, Open Solutions, 7,* 124379–124389. doi:10.1109/ACCESS.2019.2937347

Ungerer, O. (2021). FRAND in IoT ecosystems. *Intellectual Property Magazine,* (July/August), 60–61.

Urquhart, L., & McAuley, D. (2018). Avoiding the internet of insecure industrial things. *Computer Law & Security Review, 34*(3), 450–466. doi:10.1016/j.clsr.2017.12.004

Uzunov, A. V., Nepal, S., & Baruwal Chhetri, M. (2019). Proactive antifragility: A new paradigm for next-generation cyber defence at the edge. *Proceedings - 2019 IEEE 5th International Conference on Collaboration and Internet Computing, CIC 2019,* 246-255. 10.1109/CIC48465.2019.00039

Vailshery, L. S. (2021, March 8). *Global IOT and non-IoT Connections 2010-2025.* Statista. Retrieved from https://www.statista.com/statistics/1101442/iot-number-of-connected-devicesworldwide/

Van den Broeck, J., Coppens, B., & De Sutter, B. (2021). Obfuscated integration of software protections. *Int. J. Inf. Secur., 20,* 73–101. doi:10.1007/s10207-020-00494-8

Varghese, B., Reano, C., & Silla, F. (2018). Accelerator Virtualization in Fog Computing: Moving from the Cloud to the Edge. *IEEE Cloud Comput, 5*(6), 28–37. doi:10.1109/MCC.2018.064181118

Vasiliev, V. A., & Aleksandrova, S. V. (2020). The prospects for the creation of a digital quality management system DQMS. *Proceedings of the 2020 IEEE International Conference "Quality Management, Transport and Information Security, Information Technologies", IT and QM and IS 2020,* 53-55. 10.1109/ITQMIS51053.2020.9322890

Veile, J. W., Kiel, D., Müller, J. M., & Voigt, K. (2020). Lessons learned from industry 4.0 implementation in the German manufacturing industry. *Journal of Manufacturing Technology Management, 31*(5), 977–997. doi:10.1108/JMTM-08-2018-0270

Vermasan, O., & Friess, P. (2014). *Internet of Things- From Research and Innovation to Market, Deployment.* River Publishers.

Voronova, L. I., Bezumnov, D. N., & Voronov, V. I. (2019). Development of the research stand «smart city systems» INDUSTRY 4.0. *Proceedings of the 2019 IEEE International Conference Quality Management, Transport and Information Security, Information Technologies IT and QM and IS 2019*, 577-582. 10.1109/ITQMIS.2019.8928370

Wadhwa, H., & Aron, R. (2018). Fog Computing with the Integration of Internet of Things: Architecture, Applications, and Future Directions. IEEE Intl Conf on Parallel & Distributed Processing with Applications, Ubiquitous Computing & Communications, Big Data & Cloud Computing, Social Computing & Networking, Sustainable Computing & Communications (ISPA/IUCC/BDCloud/SocialCom/SustainCom), 987-994.

Wang, F., Lin, T., Tsai, H., & Lu, Y. (2014). Applying RSA signature scheme to enhance information security for RFID based power meter system. *WIT Transactions on Information and Communication Technologies*, *49*, 549-556. 10.2495/ICIE130652

Weber, R. H., & Studer, E. (2016). Cybersecurity in the internet of things: Legal aspects. *Computer Law & Security Review*, *32*(5), 715–728. doi:10.1016/j.clsr.2016.07.002

Webster, G. D., Harris, R. L., Hanif, Z. D., Hembree, B. A., Grossklags, J., & Eckert, C. (2018). Sharing is caring: Collaborative analysis and real-time enquiry for security analytics. *Proceedings - IEEE 2018 International Congress on Cybermatics: 2018 IEEE Conferences on Internet of Things, Green Computing and Communications, Cyber, Physical and Social Computing, Smart Data, Blockchain, Computer and Information Technology, iThings/GreenCom/CPSCom/SmartData/Blockchain/CIT 2018*, 1402-1409. 10.1109/Cybermatics_2018.2018.00240

Wee, H., & Wichs, D. (2021). Candidate Obfuscation via Oblivious LWE Sampling. In A. Canteaut & F. X. Standaert (Eds.), Lecture Notes in Computer Science: Vol. 12698. *Advances in Cryptology – EUROCRYPT 2021. EUROCRYPT 2021*. Springer. doi:10.1007/978-3-030-77883-5_5

Weiser, M. (1991). The computer for the 21 St Century. *Scientific American*, *265*(3), 94–105. https://www.jstor.org/stable/24938718

Wernke, M., Skvortsov, P., & Dürr, F. (2014). A classification of location privacy attacks and approaches. *Pers Ubiquit Comput*, *18*, 163–175. doi:10.1007/s00779-012-0633-z

What is the internet of things (IOT)? (2022). Retrieved from https://www.oracle.com/internet-of-things/what-is-iot/

Wireless World, R. F. (2018). *Difference between open loop power control vs closed loop power control*. Retrieved from https://www.rfwireless-world.com/Terminology/Open-Loop-Power-Control-vs-Closed-Loop-Power-Control.html

Wolff, C., & Nuseibah, A. (2017). A projectized path towards an effective industry-university-cluster: Ruhrvalley. *Proceedings of the 12th International Scientific and Technical Conference on Computer Sciences and Information Technologies, CSIT 2017*, *2*, 123-131. 10.1109/STC-CSIT.2017.8099437

Wortmann, F., & Flüchter, K. (2015). Internet of things. *Business & Information Systems Engineering*, *57*(3), 221–224. doi:10.100712599-015-0383-3

Wu, Z., Li, G., Shen, S., Lian, X., Chen, E., & Xu, G. (2021). Constructing dummy query sequences to protect location privacy and query privacy in location-based services. *World Wide Web (Bussum)*, *24*(1), 25–49. doi:10.100711280-020-00830-x

Xing, H., & Hakola, S. (2010). The investigation of power control schemes for a device-to-device communication integrated into OFDMA cellular system. In *Proceedings of 21st Annual IEEE International Symposium on Personal, Indoor and Mobile Radio Communications*. IEEE.

Xu, J., Glicksberg, B. S., Su, C., Walker, P., Bian, J., & Wang, F. (2021). Federated Learning for Healthcare Informatics. *Journal of Healthcare Informatics Research*, *5*(1), 1–19. doi:10.100741666-020-00082-4 PMID:33204939

Ye, X., Zhou, J., Li, Y., Cao, M., Chen, D., & Qin, Z. (2021). A location privacy protection scheme for convoy driving in autonomous driving era. *Peer-to-Peer Networking and Applications*, *14*(3), 1388–1400. doi:10.100712083-020-01034-w

Yousefpoor, E., Barati, H., & Barati, A. (2021). A hierarchical secure data aggregation method using the dragonfly algorithm in wireless sensor networks. *Peer-to-Peer Networking and Applications*, *14*(4), 1917–1942. doi:10.100712083-021-01116-3

Yuehong, Y. I. N., Zeng, Y., Chen, X., & Fan, Y. (2016). The internet of things in healthcare: An overview. *Journal of Industrial Information Integration*, *1*, 3–13. doi:10.1016/j.jii.2016.03.004

Yu, S. H. (2008). Methods for the Revitalization about LBS Mobile Games-Comparative Analysis between Internal and Overseas Case Study. *The Journal of the Korea Contents Association*, *8*(11), 74–84. doi:10.5392/JKCA.2008.8.11.074

Zhang, Y., Zou, J., & Guo, R. (2021). Efficient privacy-preserving authentication for V2G networks. *Peer-to-Peer Networking and Applications*, *14*, 1366–1378. https://doi.org/10.1007/s12083-020-01018-w

About the Contributors

Marcel Ohanga Odhiambo has a BSc degree in Electronic Engineering from University of Nairobi, an MSc degree in Microprocessor Technology and Applications from Brighton University and a PhD degree in Parallel and Distributed Computer Architecture from University of Surrey. He is currently Retired Research Professor, Faculty of Engineering, Mangosuthu University of Technology, Umlazi, KwaZulu-Natal, South Africa, Adjunct Professor of Electronics Engineering, Department of Electrical and Communications Engineering, Faculty of Engineering, Masinde Muliro University of Science & Technology, Kakamega, Kenya, Visiting Adjunct Professor, Department of Electrical & Communications Engineering, School of Engineering, Moi University, Eldoret, Kenya. Associate Professor January 2012 - December 2014, Department of Electrical & Mining Engineering, University of South Africa, Associate Professor & Head, Department of Process Control & Computer Systems 1st December 2014 - 31st March 2020. He is a Fellow of the Institution of Engineers and Technologists (IET), Senior Member of South African Institute of Electrical Engineers (SAIEE), Member of the Institution of Engineers of Kenya, a Registered Engineer (REng): Engineers Registration Board of Kenya (ERBK), Chartered Engineer (CEng): Engineering Council (UK) and a Professional Engineer (PrEng): Engineering Council of South Africa (ECSA). Prof. Ohanga research interest is in Parallel and Distributed Computer Architectures (Hardware), Telecommunications (Wireless Communication Networks), Electronic Systems/Devices, Intelligent Agents, Process Control/Instrumentation, Internet of Things (IoT) and Artificial Intelligence. His record spans over 66 publications in journals, conferences, book and book chapters. He has supervised 17 postgraduate students.

Weston Mwashita is with Vaal University of Technology (VUT) in South Africa. He obtained a Bachelor of Technology degree in Electrical Engineering (Telecommunications) at the University of South Africa and went on to complete his Master of Technology degree in Electrical Engineering (Telecommunications) at the same university. He obtained PhD degree from Vaal University of Technology in 2022. 18:51. He is a registered member of the South Africa Institute of Electrical

Engineers, a member of Institute of Electrical and Electronics Engineers, a member of Namibia Engineering Council, an associate member of Engineering Professions of Namibia. He has authored several publications. His research focus areas are Internet of Things, Device-to-Device Communications, Green Cellular Networks and Wireless Sensor Networks.

* * *

Ambika N. is a MCA, MPhil, Ph.D. in computer science. She completed her Ph.D. from Bharathiar university in the year 2015. She has 16 years of teaching experience and presently working for St.Francis College, Bangalore. She has guided BCA, MCA and M.Tech students in their projects. Her expertise includes wireless sensor network, Internet of things, cybersecurity. She gives guest lectures in her expertise. She is a reviewer of books, conferences (national/international), encyclopaedia and journals. She is advisory committee member of some conferences. She has many publications in National & international conferences, international books, national and international journals, and encyclopaedias. She has some patent publications (National) in computer science division.

Keerti Pendyal is an Assistant Professor and Assistant Dean (Outreach & Promotion) at the Jindal School of Banking & Finance (JSBF), O.P. Jindal Global University, India. He completed his PhD in Public Policy and Management from IIM Calcutta. He focuses on Standard Essential Patents and their policy implications in his research. He teaches courses on Management, Political Economy, and the Banking System at JSBF.

Sabyasachi Pramanik is a Professional IEEE member. He obtained a PhD in Computer Science and Engineering from the Sri Satya Sai University of Technology and Medical Sciences, Bhopal, India. Presently, he is an Assistant Professor, Department of Computer Science and Engineering, Haldia Institute of Technology, India. He has many publications in various reputed international conferences, journals, and online book chapter contributions (Indexed by SCIE, Scopus, ESCI, etc.). He is doing research in the field of Artificial Intelligence, Data Privacy, Cybersecurity, Network Security, and Machine Learning. He is also serving as the editorial board member of many international journals. He is a reviewer of journal articles from IEEE, Springer, Elsevier, Inderscience, IET, and IGI Global. He has reviewed many conference papers, has been a keynote speaker, session chair and has been a technical program committee member in many international conferences. He has authored a book on Wireless Sensor Network. Currently, he is editing 6 books from IGI Global, CRC Press, EAI/Springer and Scrivener-Wiley Publications.

Albérico Travassos Rosário, Ph.D. Marketing and Strategy of the Universities of Aveiro (UA), Minho (UM) and Beira Interior (UBI). With affiliation to the GOVCOPP research center of the University of Aveiro. Master in Marketing and Degree in Marketing, Advertising and Public Relations, degree from ISLA Campus Lisbon-European University | Laureate International Universities. Has the title of Marketing Specialist and teaches with the category of Assistant Professor at IADE-Faculty of Design, Technology and Communication of the European University and as a visiting Associate Professor at the Santarém Higher School of Management and Technology (ESGTS) of the Polytechnic Institute of Santarém. He taught at IPAM-School of Marketing | Laureate International Universities, ISLA- Higher Institute of Management and Administration of Santarém (ISLA-Santarém), was Director of the Commercial Management Course, Director of the Professional Technical Course (TeSP) of Sales and Commercial Management, Chairman of the Pedagogical Council and Member of the Technical Council and ISLA-Santarém Scientific Researcher. He is also a marketing and strategy consultant for SMEs.

Peter Ryan studied Mathematics and Physics at the University of Melbourne and earned a Bachelor of Science (Hons 1) in 1975 and a PhD in Nuclear Physics in 1981. He worked as a Research Scientist for three decades for the Australian Defence Science and Technology Organisation (DSTO) and published over 180 reports and conference papers in areas related to modelling and simulation of military operations. He currently serves on several committees of Standards Australia and in 2020 co-founded Ryan Watson Consulting Pty Ltd.

Anjum Sheikh is working as an Assistant Professor in the Department of Electronics & Communication Engineering at Rajiv Gandhi College of Engineering Research & Technology, Chandrapur. Her areas of research interest include wireless communication, the Internet of Things, Cloud Computing, and network security.

Richard Watson worked for many years as an Operations Research Scientist in the Australian Defence Science and Technology Organisation (DSTO). He has also been a Lecturer in Information Technology (IT) at Swinburne University of Technology in Melbourne and has held various roles in the IT industry in Melbourne, Canberra and Auckland, NZ. In recent years he has done research and consulting in IT and in 2020 co-founded Ryan Watson Consulting Pty Ltd to support clients in smart city and data analytics initiatives.

Index

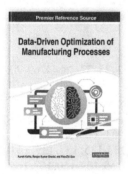

Ensure Quality Research is Introduced to the Academic Community

Become an Evaluator for IGI Global Authored Book Projects

Premier Reference Source

Stabilizing Currency and Preserving Economic Sovereignty Using the Grondona System

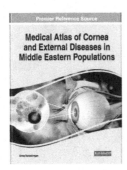

Premier Reference Source

Medical Atlas of Cornea and External Diseases in Middle Eastern Populations

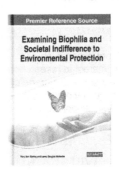

Premier Reference Source

Examining Biophilia and Societal Indifference to Environmental Protection

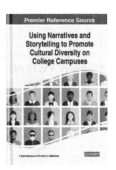

Premier Reference Source

Using Narratives and Storytelling to Promote Cultural Diversity on College Campuses

The overall success of an authored book project is dependent on quality and timely manuscript evaluations.

Applications and Inquiries may be sent to:
development@igi-global.com

Applicants must have a doctorate (or equivalent degree) as well as publishing, research, and reviewing experience. Authored Book Evaluators are appointed for one-year terms and are expected to complete at least three evaluations per term. Upon successful completion of this term, evaluators can be considered for an additional term.

If you have a colleague that may be interested in this opportunity, we encourage you to share this information with them.

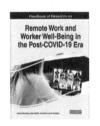

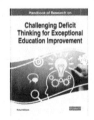

Printed in the United States
by Baker & Taylor Publisher Services